Black Holes: A Very Short Introduction

VERY SHORT INTRODUCTIONS are for anyone wanting a stimulating and accessible way into a new subject. They are written by experts, and have been translated into more than 45 different languages.

The series began in 1995, and now covers a wide variety of topics in every discipline. The VSI library now contains over 500 volumes—a Very Short Introduction to everything from Psychology and Philosophy of Science to American History and Relativity—and continues to grow in every subject area.

Titles in the series include the following:

ACCOUNTING Christopher Nobes
ADOLESCENCE Peter K. Smith
ADVERTISING Winston Fletcher
AFRICAN AMERICAN RELIGION
    Eddie S. Glaude Jr
AFRICAN HISTORY John Parker and
    Richard Rathbone
AFRICAN RELIGIONS
    Jacob K. Olupona
AGEING Nancy A. Pachana
AGNOSTICISM Robin Le Poidevin
AGRICULTURE Paul Brassley and
    Richard Soffe
ALEXANDER THE GREAT
    Hugh Bowden
ALGEBRA Peter M. Higgins
AMERICAN HISTORY Paul S. Boyer
AMERICAN IMMIGRATION
    David A. Gerber
AMERICAN LEGAL HISTORY
    G. Edward White
AMERICAN POLITICAL
    HISTORY Donald Critchlow
AMERICAN POLITICAL PARTIES
    AND ELECTIONS L. Sandy Maisel
AMERICAN POLITICS
    Richard M. Valelly
THE AMERICAN
    PRESIDENCY Charles O. Jones
THE AMERICAN REVOLUTION
    Robert J. Allison
AMERICAN SLAVERY
    Heather Andrea Williams
THE AMERICAN WEST Stephen Aron

AMERICAN WOMEN'S HISTORY
    Susan Ware
ANAESTHESIA Aidan O'Donnell
ANARCHISM Colin Ward
ANCIENT ASSYRIA Karen Radner
ANCIENT EGYPT Ian Shaw
ANCIENT EGYPTIAN ART AND
    ARCHITECTURE Christina Riggs
ANCIENT GREECE Paul Cartledge
THE ANCIENT NEAR EAST
    Amanda H. Podany
ANCIENT PHILOSOPHY Julia Annas
ANCIENT WARFARE Harry Sidebottom
ANGELS David Albert Jones
ANGLICANISM Mark Chapman
THE ANGLO-SAXON AGE
    John Blair
ANIMAL BEHAVIOUR
    Tristram D. Wyatt
THE ANIMAL KINGDOM
    Peter Holland
ANIMAL RIGHTS David DeGrazia
THE ANTARCTIC Klaus Dodds
ANTISEMITISM Steven Beller
ANXIETY Daniel Freeman and
    Jason Freeman
THE APOCRYPHAL GOSPELS
    Paul Foster
ARCHAEOLOGY Paul Bahn
ARCHITECTURE Andrew Ballantyne
ARISTOCRACY William Doyle
ARISTOTLE Jonathan Barnes
ART HISTORY Dana Arnold
ART THEORY Cynthia Freeland

Katherine Blundell

BLACK HOLES

A Very Short Introduction

OXFORD
UNIVERSITY PRESS

OXFORD

UNIVERSITY PRESS

Great Clarendon Street, Oxford, OX2 6DP,
United Kingdom

Oxford University Press is a department of the University of Oxford.
It furthers the University's objective of excellence in research, scholarship,
and education by publishing worldwide. Oxford is a registered trade mark of
Oxford University Press in the UK and in certain other countries

© Katherine Blundell 2015

The moral rights of the author have been asserted

First edition published in 2015

Published in the United States of America by Oxford University Press
198 Madison Avenue, New York, NY 10016, United States of America

British Library Cataloguing in Publication Data
Data available

Library of Congress Control Number: 2015944817

ISBN 978-0-19-960266-7

Printed and bound by
CPI Group (UK) Ltd, Croydon, CR0 4YY

to Tim & Louise Sanders, with much love

Contents

Acknowledgements

I should like to record my warm thanks to Phillip Allcock, Russell Allcock, Steven Balbus, Roger Blandford, Stephen Blundell, Stephen Justham, Tom Lancaster, Latha Menon, John Miller, and Paul Tod for many helpful comments on drafts of this book, to Stephen Blundell for preparing the diagrams, and to Steven Lee for assistance with the optical observations.

KMB

Oxford,
April 2015

List of illustrations

Chapter 1
What is a black hole?

A black hole is a region of space where the force of gravity is so strong that nothing, not even light, can travel fast enough to escape from its interior. Although they were first conceived in the fertile imaginations of theoretical physicists, black holes have now been identified in the Universe in their hundreds and accounted for in their millions. Although invisible, these objects interact with, and can thus influence, their surroundings in a way that can be highly detectable. Exactly what the nature of that interaction is depends on proximity relative to the black hole: too close and there is no escape, but further afield some dramatic and spectacular phenomena will play out.

The term 'black hole' was first mentioned in print in an article by Ann Ewing in 1964, reporting on a symposium held in Texas in 1963, although she never mentioned who coined the expression. In 1967, American physicist John Wheeler needed a shorthand for 'gravitationally completely collapsed star' and began to popularize the term, though the concept of a collapsed star was developed by fellow Americans Robert Oppenheimer and Hartland Snyder back in 1939. In fact, the mathematical foundations of the modern picture of black holes began rather earlier in 1915, with German physicist Karl Schwarzschild solving some important equations of Einstein's (known as the field equations in his General Theory of Relativity) for the case of an isolated non-rotating mass in space.

Two decades later in the UK, a little before Oppenheimer and Snyder's work, Sir Arthur Eddington had worked out some of the relevant mathematics in the context of investigating work by the Indian physicist Subrahmanyan Chandrasekhar on what happens to stars when they die. The physical implications of Eddington's calculations, namely the collapse of massive stars when they have used up all their fuel to form black holes, Eddington himself pronounced to the Royal Astronomical Society in 1935 as being 'absurd'. Despite the apparent absurdity of the notion, black holes are very much part of physical reality throughout our Galaxy and across the Universe. Further advances were made in the United States by David Finkelstein in 1958, who established the existence of a one-way surface surrounding a black hole whose significance for what we shall study in the coming chapters is immense. The existence of this surface doesn't allow light itself to break free from the powerful gravitational attraction within and is the reason why a black hole is black. To begin to understand how this behaviour might arise we need to first understand a profound feature of the physical world: there is a maximum speed at which any particle or any object can travel.

How fast is fast?

A law of the jungle is that if you want to escape a predator you need to run fast. Unless you have exceptional cunning or camouflage, you will only survive if you are swift. The maximum speed with which a mammal can escape an unpleasant situation depends on complex biochemical relationships between mass, muscle strength, and metabolism. The maximum speed with which the most rapidly travelling entity in the Universe can travel is that exhibited by particles that have no mass at all, such as particles of light (known as *photons*). This maximum speed can be given very precisely as 299,792,458 metres per second, equivalent to 186,282 miles per second, which is almost approaching a million times faster than the speed of sound in air. If I could travel at the speed of light, I would be able to travel from my home in the UK to Australia in one fourteenth of a second, barely time to blink. Light travelling from our nearest star, the Sun, takes just eight minutes to travel to us. From our outermost planet, Neptune, it's a journey time of just a few hours for a photon. We say that the Sun is eight light-minutes away from Earth and that Neptune is a few light-hours away from us. This has the interesting consequence that if the Sun stopped shining or if Neptune suddenly turned purple, no one on Earth could find out about such important information for eight minutes or a few hours respectively.

Let's now consider how fast light can travel from even more immensely distant points in space back to Earth. The Milky Way, the Galaxy in which our Solar System resides, is a few hundred thousand *light-years* across. This means that light takes a few hundred thousand years to travel from one side of the Galaxy to the other. The Fornax cluster is the nearest cluster of galaxies to the local group of galaxies (of which the Milky Way is a significant member) and is hundreds of millions of light-years away from us. Thus, an observer on a planet orbiting a star in a galaxy within the Fornax cluster looking back to Earth right now might, if equipped with appropriate instrumentation, see dinosaurs lumbering around on Earth. However, it is only the mind-boggling vastness of the Universe that makes the motion of light look sluggish and time-consuming. The role of the speed of light as a mandatory upper limit has an intriguing effect when we start to consider how to launch rockets into space.

Escape velocity

If we wish to launch a rocket into space but its launch speed is too slow then the rocket will have insufficient *kinetic energy* to break free from the Earth's gravitational field. However, if the rocket has just enough speed to escape the gravitational pull of the Earth, we say it has reached its *escape velocity*. The escape velocity of a rocket from a massive object such as a planet is larger the more massive the planet is and larger the closer the rocket is to the *centre of mass* of the planet. The escape velocity v_{esc} is written as $v_{esc} = \sqrt{2GM/R}$ where M is the mass of the planet and R is the separation of the rocket from the planet's centre of mass and G is a constant of Nature known as Newton's gravitational constant. Gravity always acts so that it pulls the rocket towards the centre of the planet or star in question, towards a point known as the *centre of mass*. However, the value of the escape velocity is completely independent of the mass of the rocket. Thus, the escape velocity of a rocket at Cape Canavaral, some 6,400 km away from the centre of mass of Planet Earth, takes the same value, just over 11 km/s or approximately 34 times the speed of sound (which may be written as Mach 34), irrespective of whether its internal payload is a few feathers or several grand pianos. Now, suppose we could shrink the entire mass of Planet Earth so that it occupies a much smaller volume. Let's say that its radius becomes one quarter of its current value. If the rocket was launched at a distance of 6,400 km away from the centre of mass, its escape velocity would remain the same. However, if it relocated to the new surface of the shrunken Earth 1,600 km from its centre, then the escape velocity would be double the original value.

Now suppose some disaster occurs with the result that the entire mass of the Earth were shrunk to a point, having no spatial extent whatsoever. We call such an object a *singularity*. It has now become a 'point mass', a massive object that occupies zero volume of space. At a very small distance of only one metre away from this singularity, the escape velocity would be much larger than it was at 1,600 km (and in fact would be about 10% of the speed of light). Closer to the singularity still, just under one centimetre away the escape velocity would be equal to the speed of light. At this distance, light itself would not have sufficient speed to escape this gravitational pull. This is the key idea to understand how black holes work.

It is worth clarifying use of the word 'singularity'. We do not believe that at the end point of a continuing gravitational collapse the matter goes down to a geometric point but rather that our classical theory of gravity breaks down and we enter a quantum regime. From here on, we will use the term singularity to refer to this ultra-dense state.

The event horizon

Now imagine you are an astronaut flying a spacecraft and that you are approaching this singularity. While still at some distance from it, you could always throw your engines into reverse and retreat from it. But the closer you get, the harder a dignified retreat becomes. Eventually you reach a distance from which it is impossible to escape, no matter how powerful your onboard engines are. This is because you have reached the *event horizon*, a mathematically-defined spherical surface, which is defined as being the boundary inside of which the escape velocity would exceed the speed of light. For our thought-experiment about Earth collapsed to a point, this surface would be a sphere of radius only one centimetre with the singularity at its centre, easy enough perhaps for our spacecraft to avoid. However, the event horizon becomes much larger when the black hole is formed from a collapsed star rather than a collapsed planet. The event horizon has an important physical consequence: if you are on that surface or inside it, the laws of physics simply won't allow you to escape because to do so you would need to break the universal speed limit. The event horizon is a mandatory level of demarcation: outside it you have freedom to determine your destiny; inside it, and your future remains unalterably locked within.

The radius of this spherical surface is named in honour of Karl Schwarzschild, who was mentioned earlier. While a soldier in World War I, Schwarzschild provided the first exact solutions of Einstein's famous field equations that underpin general relativity. The Schwarzschild radius is written as $R_S = 2GM/c^2$ where M is the mass of the black hole, G is Newton's gravitational constant, and c is the speed of light. Using this formula, the Schwarzschild radius of the Earth comes out to be just under one centimetre. Similarly, the Schwarzschild radius of the Sun is found to be 3 km, meaning that if the mass of our Sun could all be squashed into a singularity, then at just 3 km away from this point the escape velocity would be equal to the speed of light. A black hole one billion times more massive than the Sun (i.e. having a mass of 10^9 solar masses) would have a Schwarzschild radius one billion times larger (the Schwarzschild radius of a point mass that is not rotating simply scales directly with its mass). As I describe in Chapter 6, such mammoth black holes are believed to be at the centres of many galaxies.

This description of the event horizon can be reasonably thought of within Newtonian physics. Indeed, physical entities resembling black holes were imagined centuries before Einstein and others profoundly changed our understanding of space and time. The principal thinkers who imagined 'dark stars' that resemble black holes were John Michell and Pierre-Simon Laplace, starting back in the 18th century, and I will now explain what they did.

One of the remarkable things about astronomy is how much you can discover about the Universe even when you are stuck on planet Earth. For example, no human being has ever visited the Sun, and yet the presence of helium in the Sun was detected in the late 19th century by analysing the spectrum of sunlight. This is particularly remarkable as this constituted the discovery of the element helium itself; it was found on the Sun long before being detected on Earth. Even earlier, in the 18th century, some of the ideas behind black holes were beginning to be formulated, and in particular the idea of what is called a dark star. The person who made the first imaginative leap was very much a product of his time.

John Michell

The Georgian era was, in England, a time of relative peace. The English Civil War was long in the past, and England had become a land of relative domestic tranquillity (the rise of Napoleonic France was still some way off). Like his father before him, the Reverend John Michell (Figure 1) received a university education and entered the Church of England. As a rector in Thornhill, West Yorkshire, Michell was able to continue his scientific research, following up his interests in geology, magnetism, gravity, light, and astronomy. In common with other scientists working in England at the time, such as the astronomer William Herschel and the physicist Henry Cavendish (who was a personal friend), Michell was able to ride the wave of the new Newtonian thinking. Sir Isaac Newton had revolutionized the way in which the Universe was perceived, formulating his law of gravitation which explained the orbits of the planets in the Solar System as being due to the same force that caused his famous apple to drop from the tree.

1. John Michell, polymath.

Newtonian ideas allowed the Universe to be studied using mathematics, and this fresh breed of scientists was able to deploy this novel world-view into different fields. Michell was particularly concerned to use Newtonian thinking to estimate the distance to nearby stars by using measurements of the light they emitted. He came up with various schemes to do this, by relating a star's brightness to its colour; he also considered *binary stars* (pairs of stars gravitationally bound to one another) and how their orbital motions could give useful dynamical information. Michell also investigated how stars tend to cluster in particular areas of the sky, testing this against a random distribution and inferring gravitational clustering. None of these ideas was practicable at the time: few binary stars were known (though Herschel was producing some impressive catalogues of various double stars and new objects) and the relationship between a star's brightness and its colour turned out to be not quite as Michell had thought it was. Nevertheless, Michell was straining to do for the wider Universe what Newton had done for the Solar System: allow a scientific, rational, and dynamical analysis of observations to provide new information about the properties, masses, and distances of the heavenly bodies.

One particular insight that came to Michell followed from the idea that particles of light are, in Michell's words, 'attracted in the same manner as all other bodies with which we are acquainted; that is, by forces bearing the same proportion to their vis inertiae [by which he meant mass], of which there can be no reasonable doubt, gravitation being, as far as we know, or have any reason to believe, an universal law of nature'. Such particles emitted from a large star would, he reasoned, be slowed down by the gravitational attraction of the star. Thus the starlight reaching Earth would be slower. Newton had shown that light slows down in glass, and this explained the principle of refraction. If starlight was indeed similarly slowed, Michell argued that it might be possible to detect this slowing by examining starlight through a prism. The experiment was tried, not by Michell, but by the Astronomer

Royal, the Reverend Dr Nevil Maskelyne, who looked for the diminishing of the refractability of starlight. Cavendish wrote to Michell to tell him that it hadn't worked and that 'there is not much likelyhood [sic] of finding any stars whose light is sensibly diminished'. Michell was dismayed, but such astronomical speculations required much guessing of imponderables: was starlight affected by the gravitational attraction of the star from which it is emitted? Michell couldn't be sure. But he was bold enough to make an interesting prediction.

If a star was sufficiently massive, and gravity really did affect starlight, then the gravitational force could be sufficient to hold back the particles of light completely and prevent them from leaving. Such an object would be a *dark star*. This little-known cleric writing in his rectory in Yorkshire had thus been the first person to conceive of a black hole. However, so far Michell's own programme of measuring the distances to stars lay in tatters. What was more, his health had been indifferent and this had stopped him using his telescope. Cavendish wrote to him a consoling letter: 'if your health does not allow you to go on with [the telescope] I hope it may at least permit the easier and less laborious employment of weighing the world.' This singular example of a joke from Cavendish (who was notoriously buttoned up) refers to another idea that Michell had conceived. 'Weighing the world' meant an experiment in which two large lead spheres at either end of the beam of a torsion balance are attracted by two stationary lead spheres. This allows one to measure the strength of the gravitational force, and thereby infer the weight of the Earth. No one had ever done this before. Michell's idea was brilliant, but he didn't live to complete the project. Instead, Michell's experiment was performed by Cavendish and is now known as Cavendish's experiment. This transfer of credit to Cavendish is more than compensated for by the numerous breakthroughs made by Cavendish which he neglected to publish and were later attributed to subsequent researchers (including 'Ohm's' law and 'Coulomb's' law).

Pierre-Simon Laplace

On the other side of the English Channel, Pierre-Simon Laplace did not enjoy the tranquil idyll afforded by the peaceful period of the English Enlightenment. Laplace lived through the French Revolution, though his career prospered as he influenced the newly founded Institut de France and the École Polytechnique. He even spent a period as Minister for the Interior under Napoleon, a short-lived appointment the Emperor came to regret. Napoleon realized that Laplace was a first-rate mathematician but as an administrator he was worse than average. Napoleon later wrote of Laplace that 'he sought subtleties everywhere, conceived only problems, and finally carried the spirit of "infinitesimals" into the administration'. Napoleon had other administrators to call upon, but the world has had few mathematicians as productive and insightful as Laplace. He made pivotal contributions to geometry, probability, mathematics, celestial mechanics, astronomy, and physics. He worked on topics as diverse as capillary action, comets, inductive reasoning, solar system stability, the speed of sound, differential equations, and spherical harmonics. One of the ideas he considered was dark stars.

In 1796 Laplace published his *Exposition du système du monde*. Written for an educated public, this book describes the physical principles on which astronomy is based, the law of gravity and the motion of the planets in the Solar System, and the laws of motion and mechanics. These ideas are applied to various phenomena, including the tides and the precession of the equinoxes, and the book also contains Laplace's speculations on the origin of the Solar System. One particular passage is of special relevance to our story. Laplace made a calculation of how large an Earth-like body would need to be so that its escape velocity was equal to that of light. He showed, quite correctly, that the gravitational strength on the surface of a star, with density comparable to that of Earth but with a diameter of about 250 times that of the Sun, would be so intense that not even light would be able to escape. Thus, he reasoned, the largest bodies in the Universe would therefore be invisible. Could they still be lurking, undetectable in the dark night sky, while we imagined that the only things 'out there' were the bright luminous objects that we can see? The Hungarian astronomer Franz Xaver von Zach requested that Laplace provide the calculations that led to this conclusion, and Laplace obliged, writing this up (in German) for one of the journals that von Zach edited.

However, Laplace was becoming aware of the wave theory of light. Both Michell's and Laplace's ideas were based in part on the corpuscular theory of light. If light were to consist of tiny particles, then it seemed reasonable that these particles would be affected by a gravitational field and would be bound forever to a star of sufficient size. But the early 19th century saw a number of experiments which seemed to give greater credence to the wave theory of light. If light were instead a wave, then it was harder to see that it should be affected by gravity. Laplace's dark star prediction was quietly omitted from later editions of *Exposition du système du monde*. After all, Michell and Laplace had been conjecturing and exploring theory, rather than being driven by the need to explain observations and thus, this idea was forgotten for a while. The objects imagined by Michell and Laplace were thus 'dark stars', enormous objects in the Universe which by virtue of their mass could sustain planetary systems but by virtue of this same overwhelming bulk could not be observed via the radiation of light. Starlight emitted from the surfaces of Michell's and Laplace's dark stars would be too sluggish to overcome the intense surface gravity. What Michell and Laplace could not have guessed was that such gargantuan accumulations of mass would be unstable to collapse. Moreover, in their collapse they would puncture the very fabric of space and time and give rise to a singularity. Thus 'black holes' are not 'dark stars' and to take the argument forward and begin to meet up with the astronomical discovery of black holes we will first need to understand the nature of spacetime.

Spacetime

Our everyday experience leaves us comfortable with the notion that the tangible Universe may be described by one temporal (or time) coordinate t and three spatial coordinates (for example x, y, and z along three mutually perpendicular axes, a construct invented by René Descartes and known as Cartesian coordinates). In 1905, Einstein published his revolutionary paper on Special Relativity, the relativity of motion and stationarity. In 1907, Hermann Minkowski showed how these results could be understood more deeply by considering a four-dimensional spacetime whose points, specified now by the 4-D coordinate (t, x, y, z), correspond to 'events'. An event is something that happens at a particular time (t) and at a particular place (x, y, z). Such 4-D coordinates in what is known as Minkowski spacetime specify exactly where and when an event occurs. Einstein's special theory of relativity could be formulated in terms of Minkowski's spacetime and provides a convenient description of physical processes in different frames of reference that move relative to one another. A 'frame of reference' is simply the perspective possessed by a particular observer. Einstein called this theory 'special' because it deals only with a particular case, namely reference frames that are non-accelerating (called inertial frames of reference). The special theory can only be applied to uniformly moving, non-accelerated, frames of reference. If you drop a stone, it accelerates towards the ground. The frame of reference attached to the stone is an accelerating frame of reference and cannot be treated by Einstein's special theory. Where you have gravity, you have acceleration.

This drawback prompted Einstein to formulate a *general* theory of relativity, which he published a decade after his special theory. What he found was that whereas Cartesian space and Minkowski spacetime were rigid frameworks in which objects 'live, move and have their being', spacetime was actually a more responsive entity: it could be curved and otherwise deformed by the presence of mass. Once mass is present in a physical situation then the following inextricably linked behaviour describes reality, neatly summarized by John Wheeler:

- mass acts on spacetime, telling it how to curve
- spacetime acts on mass, telling it how to move

This behaviour is quantified by Einstein's field equations within General Relativity, which relate the curvature of spacetime to the gravitational field.

Physicists talk about a *gravitational potential well* as surrounding a massive object. The cartoon shown in Figure 2 encapsulates how the spacetime is distorted in the vicinity of a couple of black holes, where each region can be regarded as curved in a way which is directly related to its mass and hence to the gravitational force itself. The singularity in spacetime may be regarded as where the curvature in spacetime becomes very high and you go beyond the classical theory of gravity, into the quantum regime. The event horizon surrounding the singularity functions as a one-way

2. **The distortion, i.e. curvature, in spacetime due to the presence of masses.**

membrane: particles and photons can enter the black hole from outside but nothing can escape from within the horizon of the black hole out to the external Universe. In fact mass is not the only property that a black hole may possess and be measured by. If the black hole is rotating, that is to say it possesses some spin, then even more extreme behaviour emerges. Before we examine this, we will take a little detour to learn a little more about how we may schematically represent spacetime itself.

Chapter 2
Navigating through spacetime

Mathematics is the exquisitely perfect language needed for describing how the theory of relativity applies to the physical Universe and all of spacetime, and that description includes the strange behaviour that occurs near black holes. A mathematical description, while powerful and exact, even so can be something of a foreign and forbidding language for those without the appropriate technical training. Descriptive words, however eloquent, lack the rigour and power of a mathematical equation and can be imprecise and limiting. Pictures however, being (it is said) worth a thousand words, can be not only a useful compromise but a very helpful way to visualize what is going on. For this reason, it is well worth spending a little effort to understand a particular type of picture, called a spacetime diagram. This will help in understanding the nature of spacetime around black holes.

Spacetime diagrams

The cartoon in Figure 3 shows a simple spacetime diagram. Following tradition, the 'time-like' axis is the one that is vertical on the page and the 'space-like' axis is drawn perpendicular to this. Of course, we really need four axes to describe spacetime because there are three space-like axes (usually denoted x, y, and z) and one time-like axis. However, two axes will suffice for our purpose

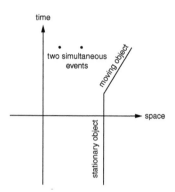

3. A simple spacetime diagram.

(and of course four mutually perpendicular axes are impossible to draw!). Where these two axes intersect is called the origin, and this may be regarded as the point of 'here and now' for the observer who has constructed their spacetime diagram. An idealized instantaneous event, say the click of a camera shutter, occurs at a particular moment in time and at a particular location in space. Such an instantaneous event is represented by a dot on a spacetime diagram, appropriate to the time and spatial location in question. There are two dots in Figure 3, which are spatially separated (they do not occur at the same point on the space axis) but they are simultaneous (they have the identical coordinate on the time axis). You could imagine these two dots correspond to the simultaneous shutter presses of two photographers who are standing some distance apart from one another, photographing the same spectacle. If points represent events, what do lines in a spacetime diagram represent? A line simply shows a path of an object through spacetime. As we live our lives, we journey through spacetime and the path we leave behind us (somewhat as a snail leaves a glistening trail of slime behind it) is a line in spacetime, and in the jargon this is called a *worldline*. If you stay at home all day, your worldline is a vertical path through spacetime (with space coordinate = '22 Acacia Avenue', for example). You move forward in time but are fixed in space. If on the other hand you made a long journey, your worldline slants over because your distance changes with time, because you move in space as well as time.

For example, look at the worldline shown in Figure 3, the line which is part vertical, then further up becomes slanting. This corresponds to the worldline of some other entity, which is stationary for the time indicated by the vertical extent of the line. An example might be a camera belonging to one of the photographers, left on a chair (so that its worldline is vertical because its position isn't changing), before it was stolen and whisked away (when the spatial location changes continuously). Where this line becomes slanting is where its spatial location is changing with time. The slope of this line tells you about the rate of change of distance with time, which is more commonly called the speed. In this case this is the speed at which the thief is whisking away the stolen camera. The faster the thief is making off with the camera, in other words the more ground he is covering in a given time, the less vertical and the more slanting this part of the line will be. There is of course a robust upper limit to the speed at which the thief can run off with his illegally gotten gains and this, as discussed in Chapter 1, is the speed of light. The trajectory of a beam of light would be represented by a maximally slanting line (commonly represented in spacetime diagrams as being at 45 degrees to the time axis by using cleverly crafted units). Because nothing can go faster than that speed, no worldline can be at a greater angle to the time axis than this.

Worldlines on a spacetime diagram having this maximally slanting angle, corresponding to this maximal speed, the speed of light, give rise to an important concept called a light cone. The idea of this is very simple: you can only have an effect on the Universe in the future by some prior cause and that causal sequence cannot propagate faster than the speed of light. Therefore your 'sphere of influence' right now is contained in a restricted range of

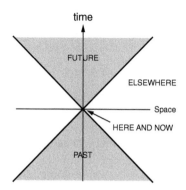

time

FUTURE

ELSEWHERE

Space

HERE AND NOW

PAST

4. A simple light cone diagram.

spacetime, namely that part which is within a 45-degree angle to the positive time axis as shown in Figure 4. Moreover, you can only have been influenced by a causal chain of events that could not have propagated faster than the speed of light. Therefore only events within a 45-degree angle to the backwards time axis can influence you now. If we now draw a spacetime diagram with two space-like axes and one time-like axis, then the triangles in Figure 4 become cones and these are what we mean by *light cones*, as shown in Figure 5. The light cone in Figure 5 delineates regions of space within which an observer (deemed to be located at the origin, their 'here and now') could in principle reach (or have reached in the past) without having to invoke breaking the cosmic speed limit and travelling faster than the speed of light. The region centred on the positive (future) time axis is known as the future light cone while the cone centred on the negative time axis (i.e. past times) is known as the past light cone.

Thus the assassination of Julius Caesar in 44 BC is part of your past, because there is a conceivable causal link between that event and you. (If you had to learn about it at school, that demonstrates the existence of a causal link!) Because light from the Andromeda Galaxy can reach a telescope on Earth, it too is part of your past. However, the light takes 6 million years to get to us, so it is the

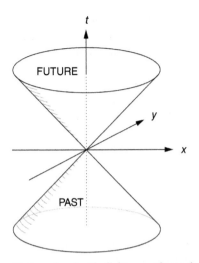

5. A spacetime diagram showing the light cone of a particular observer.

Andromeda Galaxy of 6 million years ago that is part of your past and sits on your light cone. The Andromeda Galaxy of today, or even the Andromeda Galaxy of 44 BC, is outside your light cone. Events happening on Andromeda, either now or even back in 44 BC, cannot influence you right now because any causal link would have had to travel faster than the speed of light.

The three spacetime diagrams that we have seen in this chapter so far have their axes labelled as time and space. In fact, professionals wouldn't normally include axis labels or even the axes in spacetime diagrams. This isn't simply that it is so routine that time goes up and space goes across that professional astrophysicists get sloppy (though that's not an unknown phenomenon) but it is because the exact positions in spacetime cannot be agreed upon by all observers. In the world of special relativity, the notion of simultaneity breaks down. Just because two events are seen to be simultaneous for one observer doesn't at all mean that they are simultaneous for other observers.

Thus the two photographers pressing the shutters of their cameras 'simultaneously' will not be what an observer travelling in a spacecraft very fast relative to the cameras sees. That observer will deduce one camera click occurring substantially before the other. The two points in Figure 3 which I drew at the same vertical height (since I claimed the events occurred at the same time) would appear at different vertical positions on the spacetime diagram of the rapidly travelling observer. Einstein's relativity insists her diagram is just as valid as mine. So if the points on a spacetime diagram depend on an observer's point of view, i.e. their frame of reference, what's the reason for drawing them?

To understand this, it is helpful to focus on the worldline of a moving particle and so we will now draw a new spacetime diagram in which a particle moves through spacetime, taking its light cone with it (this trick is known as working within the co-moving frame). Notice that in Figure 6 the particle's path (i.e. its worldline) always stays within the light cone as it cannot travel faster than the speed of light.

Einstein's Special Theory of Relativity, which is a subset of his General Theory, pertains to a restricted set of physical situations. A different conceptual framework beyond Special Relativity is needed in the context of spacetime which is expanding, the pre-eminent example of which is the expanding Universe. In this context, the manifestation of causality is such that you cannot move faster than the speed of light with respect to your local bit of space.

How do objects know where to go?

Although photons have no mass, it turns out that they are still influenced by gravity. It is best not to think of this as due to a force, but rather that this comes about because of the curvature of spacetime. A photon is usually thought to travel in a straight line, which is where we get the notion of a 'light ray'. However, through a curved spacetime it will follow a path known as a *geodesic*.

Worldline

6. A spacetime diagram of a particle moving along its worldline, that is always contained within its future light cone.

Despite its Earth-based connotations, a geodesic (whose name comes from geodesy, i.e. measuring the lie of the land of our planet's surface) is an important concept describing the nature of spacetime throughout the Universe. If space were not curved (meaning entirely consistent with everyday geometry that we may have learned at school from Euclid or one of his successors), then a

21

geodesic would be the 'straight line path' that a light ray would travel. But the shortest distance between two points, which is the route that a light ray 'wants' to take, is known by the term 'null geodesic'. In curved space the shortest distance between two points isn't what we think of as straight, but 'geodesics are straight lines in curved spaces'. A straight line can also be characterized as the path you follow by keeping moving in the same direction. An example of how geometry is seriously different on a curved surface comes from considering lines of longitude on a sphere. Two adjacent lines of longitude (which are parallel to one another at the equator) will meet at a point at the pole, as shown in Figure 7. However, in flat space parallel lines will meet only at infinity (as per Euclid's last axiom).

Actually, where spacetime is curved, for example because of the presence of mass, that curvature is manifested in the path that a light ray or (a mental device used by physicists) a 'test particle' freely able to move with no influence of any external force, would

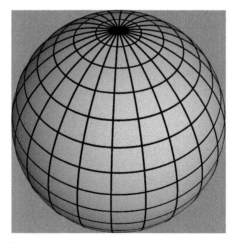

7. Lines of longitude on a sphere are parallel at the equator, and meet at a point at the poles.

move along between two events. Two events should be regarded as two points in 4-D space time, each denoted in the form (t, x, y, z).

A rule called a *metric* tells us how clocks and rulers measure the separations between events in space and time and provide the basis for working out problems in geometry. A very simple example of a metric is Pythagoras' theorem, which tells us how to compute the distance between two points that lie in a plane. The solutions to Einstein's field equations tell us how to calculate the metric of spacetime when the distribution of matter is known. We use this to construct the geodesics for the real Universe. For example, one of the first pieces of observational evidence for General Relativity was the bending of starlight by the Sun, measured during a solar eclipse (a good time to examine the apparent positions of stars close to the Sun's disc because light from the disc is blocked out by the Moon, an opportunity seized upon by Sir Arthur Eddington in 1919). The Sun's mass curves spacetime. Thus the shortest path (the geodesic) from a distant star to a telescope on Earth is not quite a straight line: it is bent round by the Sun's gravitational field, as shown in Figure 8.

The bending of starlight demonstrates that space is curved, but Einstein's General Theory tells us it is actually spacetime that is curved. Therefore we might expect that mass also has some

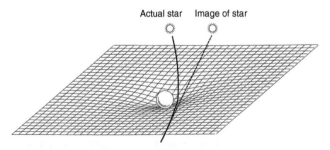

Actual star   Image of star

8. A mass such as the Sun causes distortion, or curvature, in spacetime.

strange effects on time. In fact, even the Earth's gravitational field is sufficient to make Earth-bound clocks tick a bit slower than they would do in deep space, although the effect is small (roughly one part in a billion) but measurable. The gravitational effects near the event horizon of a black hole are much stronger. Thus, even for the simplest case of a non-spinning black hole, time runs differently close to the black hole compared to how it runs at a huge distance from the black hole. This is a real effect and does not depend on how the time is measured (for example by an atomic clock, or by a digital watch). It follows directly from the curvature of spacetime induced by the mass which tips the light cones towards the mass. Figure 9 indicates the general effect.

Black holes profoundly affect the orientations of the light cones. As a particle approaches a black hole, its future light cone tilts

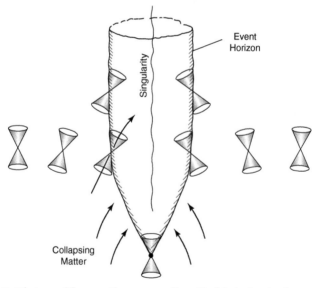

9. Diagram of the spacetime surrounding a black hole showing how the future light cones for objects on the event horizon lie inside the event horizon.

more and more towards the black hole, so that the black hole becomes more and more a part of its inevitable future. When the particle crosses the event horizon, all of its possible future trajectories end inside the black hole. Just within the event horizon, the light cone tilting is so great that one side becomes parallel with the event horizon and the future lies entirely within the event horizon; escape from the black hole is not possible. Figure 9 also illustrates this point: it is essentially a representation of 'local spacetime diagrams', because the assembly of light cones allows you to understand the local conditions experienced by a test particle located at different positions. In this figure, time increases up the page and so this diagram also gives a sense of how a black hole forms and grows due to infalling matter.

Just as for the dark stars of Michell and Laplace discussed in Chapter 1 which could have sustained planetary systems in orbit around them much like our Solar System, so it is that we only know that a black hole is nearby due to its gravitational pull. This might lead you to think that the only property that characterizes a black hole is its mass. In fact, whether or not a black hole is rotating has a dramatic effect on its properties, and I will explain how this comes about in Chapter 3.

Chapter 3
Characterizing black holes

In Chapter 1, we introduced the concept of a mass singularity, forming in gravitational collapse, and surrounded by an event horizon. Examples of such objects that are not spinning are called *Schwarzschild* black holes and this term specifically denotes black holes that are not rotating: in the jargon, they have no *spin*. Simply put, the only characteristic that distinguishes one Schwarzschild black hole from another (other than location) is how massive it is. In Chapter 7 we will learn how black holes grow but for now, it will suffice to know that collapse under gravity is the key ingredient. If there is any rotation whatsoever in the pre-collapsed matter, however gentle, then as the collapse occurs the rotation rate will increase (unless something acts to stop that happening). This arises due to a remarkable physical law known as the *conservation of angular momentum*. This law is illustrated by a pirouetting skater: as she pulls her arms in she spins faster. In the same way, if the star that gives rise to the black hole is gently rotating then the black hole that it ultimately forms will be spinning significantly and is termed a *Kerr black hole*. Most stars are in fact rotating, because they themselves are formed from the gravitational collapse of slowly rotating massive gas clouds. (If such a gas cloud had even a minute amount of net rotation then the collapsing cloud will have non-zero angular momentum, and as the matter occupies an increasingly smaller volume the final rotation of the collapsed object may well be rather rapid.) Thus we see that rotation, more commonly called spin, is likely to be a prevalent, if not actually a ubiquitous, characteristic for black holes that have just formed from the collapse of matter. We now believe that spin is as inevitable in real astrophysical black holes as it is in current-day politics (though in the latter case it arises from something other than the conservation of angular momentum!).

We have now stated that a second physical parameter, that of spin or angular momentum, is a characteristic that distinguishes one black hole from another just as mass does. Thus, there are two properties of black holes that are important to keep in mind as we study the behaviour of black holes: mass and spin. In principle, there is a third characteristic of black holes that might be relevant to their behaviour: electrical charge. This is also a conserved quantity in physics, and the forces between electric charges, known as electrostatic forces, have a number of resemblances to gravitational force. A key similarity is that both are (on large scales) examples of inverse-square laws meaning that, in the case of two massive objects, as you double the distance that separates them from one another the gravitational force they experience reduces to a quarter of the original value. A key difference is that while gravity is always attractive, electrostatic charges are only sometimes attractive (when the two bodies are oppositely charged, i.e. one is positive and the other is negative). They are at other times repulsive (when the bodies have charge of the same sign, either both positive or both negative, they repel each other). If two charged bodies have the same type of charge, then electrostatic repulsion will tend to prevent them coalescing, even if gravity is tending to attract them. So while charge could in principle be a third property of black holes that one might hope to measure, in reality a charged black hole would be rapidly neutralized by the surrounding matter. It is therefore a good operational assumption that there are only two relevant properties of black holes that distinguish one from another: mass and spin. That's all!

Now, you might wonder whether black holes could be distinguished by their composition. One might have been formed from a hydrogen gas cloud, another from a helium gas cloud. Why should it be that the provenance of the collapsed matter that gave rise to the black hole isn't manifested in the measurable properties of the black hole subsequently formed? That's because information can't get out of the event horizon! Light is the means by which information might be transmitted, but we have already seen in Chapter 1 that it cannot escape from inside the event horizon of a black hole. Thus the chemical composition of the matter that fell into the black hole can have no effect on the properties of the black hole as determined from the outside. It would not be correct to think of gravity as something that needs to 'get out of' the black hole. The continued existence of a gravitational field external to the black hole is something that is laid down in the formation of the black hole as spacetime becomes distorted. No influence from inside the black hole could change the external field after the event horizon has formed.

Black holes have no hair

When we are asked to describe another person, a distinguishing characteristic that is often included is their hair (for example, strawberry blonde, or silver grey or chocolate brown). There are sometimes clues in the nature of people's hair as to their age or their nationality. Information about further physical characteristics such as 'Body Mass Index' might provide information on their diet. In contrast to humans, black holes are entities that have absolutely no distinguishing characteristics other than their mass and their spin (neglecting charge for the reasons noted above). This is captured in the breviloquent phrase 'Black holes have no hair', coined by John Wheeler to emphasize that there is nothing about a black hole that bears any evidence of the nature of its progenitor star. Not its shape, not its lumpiness, not its landscape, not its magnetism, not its chemical composition. Nothing. Calculations done by, amongst others, the Belarusian physicist Yakov Zel'dovich demonstrated that if a non-spherical star with a lumpy surface collapsed to form a black hole, its event horizon would ultimately settle down to a smooth equilibrium shape having no lumps or bumps of any kind. So, a black hole never has a bad hair day! The only things you can know about it are its mass and spin.

Spin changes reality

Perhaps the most remarkable feature of a spinning black hole is that the gravitational field pulls objects around the black hole's axis of rotation, not merely in towards its centre. This effect is called *frame dragging*. A particle dropped radially onto a Kerr black hole will acquire non-radial (i.e. rotating) components of motion as it falls freely in the black hole's gravitational field.

What this means for a test particle having spin (such as a small gyroscope) is that if it falls freely towards a rotating massive body, such as a Kerr black hole, it will acquire a change to its spin axis. It is as though its local frame of reference was dragged by the rotation of the central massive body. This phenomenon, discovered in 1918, called the Lense–Thirring effect actually occurs not just around black holes, but to some extent around any spinning object. If you put a very precise gyroscope in orbit around the Earth, the frame dragging causes the gyroscope to precess.

It is Einstein's field equations that describe the mathematics of black holes and, as also mentioned in Chapter 1, Karl Schwarzschild solved these equations for the case of the stationary (non-rotating) black hole, a remarkable achievement given that he did this in 1915, the same year that Einstein introduced his general theory of relativity. The case of the spinning black hole was treated much later by New Zealander Roy Kerr in 1965. A few years after this, the Australian Brandon Carter explored Kerr's solution further still. Carter carried out an in-depth investigation into the consequences of the Kerr metric. He established that a spinning black hole causes a dramatic swirling vortex in the spacetime that surrounds it which arises because of the dragging of the reference frame. An example of a vortex is a whirlwind—close to the centre of the whirlwind the air swirls rapidly, carrying with it anything in its path, be it hay in a hay field or sand in a desert. Further from the whirlwind the air (and hence hay or sand) rotates much more slowly. So it is too, with spacetime surrounding a spinning black hole: far away from the event horizon, the speed at which spacetime itself rotates is slow, but at the horizon, spacetime itself spins with the same speed that the horizon spins.

The event horizon for the spinning (Kerr) black hole is much the same as for a non-spinning (Schwarzschild) black hole, except that the faster the black hole is spinning, the deeper the gravitational potential well: a Kerr black hole forms a deeper gravitational potential well than a Schwarzschild black hole of the same mass, and therefore a Kerr black hole can be a more powerful energy source than a non-spinning one, a point to which we return in Chapter 7. In the meantime, it is helpful to summarize this behaviour by saying the event horizon of a Schwarzschild black hole depends only on mass, but that of a Kerr black hole depends on both mass and spin.

An outstanding question is whether there could be, even in principle, any spacetime singularities that are not enclosed within and hidden by event horizons—a so-called 'naked singularity'. By definition, all black hole solutions to the Einstein field equations do have event horizons and, as shown in Chapter 1, no light and therefore no information can escape from within such horizons. All black hole singularities are believed to be enclosed within event horizons and therefore not 'naked', so that direct information about the singularity is inaccessible from the rest of the Universe. The so-called cosmic censorship conjecture was formulated by the British mathematician Roger Penrose and states that all spacetime singularities formed from regular initial conditions are hidden by event horizons and that there are no naked singularities out in space.

How much spin is too much?

There is a limit to how much angular momentum a black hole can have. This limit depends on the mass of the black hole, so that a more massive black hole can spin faster than a less massive black hole. A black hole that is rotating close to this maximum limit is known as an extreme Kerr black hole. It is possible to show that if you try to spin up a black hole, to make an extreme Kerr black hole, by firing rapidly rotating matter into it (i.e. giving it a stir) then centrifugal forces prevent the matter from even entering the event horizon.

Somewhat further out from the event horizon of a rotating black hole is another significant mathematical surface which is known as the *static limit*. The dragging of inertial frames means that if the spin of the massive body is non-zero then there are no stationary observers inside of this surface: every physically realizable reference frame inside the static limit must rotate. Within this surface, space is spinning so fast that light itself has to rotate with the black hole, i.e. it is impossible to remain motionless. The region between the static limit and the event horizon is known as the *ergosphere*, which rather confusingly is not spherical, as shown in Figure 10. In equatorial directions the ergosphere is much larger than the event horizon, but in the polar directions the radius of the ergosphere is the same as the radius of the event horizon. The resulting shape of the ergosphere is an oblate spheroid, resembling the shape of a Jarrahdale pumpkin (without the stalk). The first two syllables of ergosphere, however, come from the Greek noun *érgon* relating to 'work' or 'energy' (as in 'ergonomics') from which the old unit of energy, the erg, is also derived. It is intriguing to note that in addition there is a Greek verb *ergo* which means to enclose and keep away, appropriately for the nature of the ergosphere. Perhaps this may have been in the

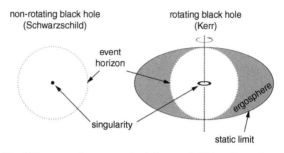

non-rotating black hole
(Schwarzschild)

rotating black hole
(Kerr)

event horizon

singularity

ergosphere

static limit

10. The different surfaces around a Schwarzschild (stationary) black hole and around a Kerr (spinning) black hole (in the frequently used representation of 'Boyer-Lindquist' coordinates).

minds of Roger Penrose and Demetrios Christodoulou who coined and championed the name of this region around a spinning black hole. The importance of the ergosphere is that it is the region within which energy can be extracted away from the black hole.

Since inside the ergosphere space is spinning, particles of matter within that space also get swept up into a rotational motion. Considerable rotational energy is therefore stored in this rotation of space, a very important point to which we return in Chapter 8.

White holes and wormholes

Einstein's equations of General Relativity are particularly rich and allow many different solutions describing alternative versions of curved spacetime. This provides an almost inexhaustible source of possible universes for cosmologists to describe and think about. Which type of universe we actually live in is a matter that can only be decided by observation (if at all!). But that doesn't stop mathematical physicists playing around with Einstein's equations to find all kinds of interesting solutions.

One intriguing object that can be dreamt up by mathematical physicists is what is called a *white hole*. A white hole behaves just like a black hole but with the direction of time reversed (imagine a movie played backwards). Instead of matter being sucked in, it is spewed out. Instead of the event horizon marking out the region from which you can never escape, it stakes out the region into which nothing could ever enter. Once matter exits from a white hole, it can never return there; its entire future is outside. As we see in Chapter 6, a black hole is formed from a collapsing star and must eventually evaporate by the laws of quantum mechanics into Hawking radiation (see Chapter 5). A white hole, on the other hand, could only result from radiation that for some reason spontaneously assembles into a black hole. It is not easy to understand how this could happen in practice, and moreover Douglas Eardley has demonstrated that white holes are inherently unstable.

When Einstein and his student Nathan Rosen were playing around with Einstein's equations in the 1930s, they found an interesting solution. If a region of spacetime could be strongly curved, it might be possible for it to become sufficiently folded that two parts of spacetime which had previously been separated by a large distance could become connected by a small bridge, or wormhole, as shown in Figure 11. The enormous distances between the stars and galaxies have always been unfavourable for those writers who wish to set human dramas on a cosmic stage, and wormholes (also known as Einstein–Rosen bridges) have provided the perfect plotting device for writers to transport their heroes and villains about. This mathematical invention has been an absolute boon to the writers of science fiction, because it provides a ready means for traversing enormous distances through space and thereby to sustain various highly artificial and unbelievable plot devices. Yet again, we have no observational evidence that wormholes actually exist in our Universe. In addition, there is considerable theoretical evidence that a wormhole, once formed, would not be stable for very long. It seems that to keep a wormhole propped open, one needs a large amount of negative energy matter, and all normal matter has

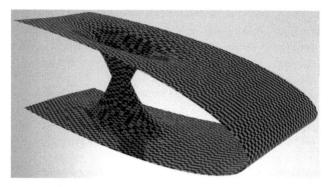

11. A wormhole connecting two otherwise separate regions of spacetime.

positive energy (this is connected with the fact that gravity is normally always attractive). Normal matter passing through a wormhole may be enough to destabilize and destroy it, causing it to turn into a black hole singularity.

If wormholes did exist, and could be maintained for any reasonable length of time, they would have some surprising and bizarre properties. Not only would they provide a means for taking an enormous shortcut across a vast expanse of space, but they would also allow a traveller to journey back in time. One would then be able to construct closed time-like curves, loops in spacetime in which the light cones form a ring (see Figure 12) so that, like in the movie *Groundhog Day*, a person travelling along a closed time-like curve would simply repeat their same experiences over and over again.

In fact, there are a number of solutions to Einstein's equations in addition to wormholes which have this alarming and counterintuitive property. In 1949, the mathematician Kurt Gödel found a solution that described a spinning universe, and this contains exactly the same sort of closed time-like curves which pass through events again and again in an endless *Groundhog Day*

12. A closed time-like loop, on which your future becomes your past.

cycle. (Evidently 'free will' is not part of the field equations!) The part of the Kerr solution thought to have genuine physical significance in the real world is that which describes the spacetime outside of the event horizon. However, it is unclear whether the part of the Kerr solution inside the event horizon, while mathematically sound, has any physical relevance. In this part of the Kerr solution, the singularity is not a point (as it is for the non-rotating black hole) but has the form of a rapidly rotating ring (however, the physical validity is very speculative). This ring-like singularity is surrounded by closed time-like curves. On such a curve, your future is also in your past and you have the theoretical possibility of murdering one of your own grandparents before they had produced your parents! Thus the existence of closed time-like curves seems to create the possibility of all kinds of paradoxes relating to time travel. One possible solution to this is to admit that we do not have a theory that links quantum mechanics (which describes the very small) and general relativity (which describes the very massive), in other words a theory of quantum gravity. We don't know the physics of extremely massive but very small objects. Most physicists think we need this to fully understand the behaviour of spacetime very close to singularities. Thus it may

35

be that these strange solutions to Einstein's equations do not actually occur in the Universe because they are prohibited by its fundamental quantum mechanical nature. Quantum effects may, for example, destabilize wormholes. Stephen Hawking believes this to be the case and has called this principle the 'Chronology Protection Conjecture'. He has quipped that this is the underlying principle that keeps the Universe safe for historians.

There is much about the interior of rotating black holes that pushes our understanding of fundamental physics to the limits and therefore to where much of our description is highly speculative. By contrast, the rotation of black holes and their effect on their surroundings is something that has enormous practical significance for understanding what we can see with our telescopes. Thus our next step is to consider in more detail what happens to matter when it falls into a black hole.

Chapter 4
Falling into a black hole . . .

How close is too close?

Before we can consider in detail what would happen if you or your belongings had the misfortune to fall into a black hole, it is important to understand the effect of an observer's particular perspective, or frame of reference. This means that different observers see very different things. Exactly what your perspective is on an object falling into a black hole depends on how far away you are from that object (and indeed whether you are that object!). Consider a particle of light, a photon, that is outside the event horizon of a black hole: since it is outside the horizon, it can in principle escape. Inside the event horizon it would be a different story—the photon could not escape the gravitational field of the black hole. But even outside the event horizon, a photon that is travelling away from the black hole will not escape completely unscathed. The photon suffers a loss in its energy due to the work it has to do against gravity. This is an example of a gravitational potential well; just as energy would be needed to haul yourself upwards out of a deep well, so the photon needs to expend energy to pull itself away from the region near a massive object. The effect has even been measured for photons moving in the Earth's gravity. The energy of a photon is inversely related to its wavelength: a high-energy photon has a short wavelength whereas a low-energy photon has a long wavelength. The photon loses energy as it retreats away from the black hole, so its wavelength increases. This changes the colour of the light, moving from the blue (short wavelength) towards the red (long wavelength) end of the spectrum (this effect is called *redshift*). This sort of redshift, known as *gravitational redshift*, arises where spacetime itself stretches out, or is curved, for example by the effect of a massive body such as a black hole. Note that John Michell, despite having significant original thoughts about dark stars, was incorrect in thinking that the velocity of light decreases as it climbs out of the potential well. We now know that it is the wavelength (hence frequency) of light that is affected by the presence of a massive star.

What happens to time near a black hole?

In Chapters 1 and 2 I described how spacetime is distorted by the presence of a mass (i.e. something which produces its own gravitational field) and this means that not just space, but also time is affected close to a black hole.

Imagine you want to keep a safe distance from a Schwarzschild black hole but you want to learn more about how time behaves nearby. Thus you have arranged for twenty-six fixed observers to be stationed close to the black hole's event horizon but definitely safely outside it. These observers are named A to Z, and are arranged in a line with A closest to the event horizon and with Z being nearest to you, safely far away. Each observer from A to Z has a good clock with which to measure their local time, at their particular location. As part of the deal to persuade A to Z to participate in this experiment, you had offered them each an inducement in the form of a gift of an additional, unusual clock that had been adjusted so that the time on it would read the same as the time on your clock at your safe distance. Participant Z, closest to you, would find that the two clocks in his possession read slightly different times because his own clock, which measures local time (*'proper time'* in the jargon), would be running slightly more slowly than the gift clock which matches the time you measure at your rather safer and more remote distance. The collated results of participants Z to A would display a remarkable effect: closer to a black hole, a clock measuring time 'runs more slowly' compared with the distant time as reported on the participants' specially adjusted gift clocks. This effect, described by Einstein's theory of general relativity, is known as *time dilation*. The effect would be greater and greater for the observers nearer the start of the alphabet who are nearer to the black hole. The greater the proximity to a black hole, the more slowly a local clock (of any kind: atomic, biochemical) will run compared to a clock used by a distant observer.

Suppose you were multi-tasking your experiments with a different set of twenty-six observers at the same distances from a different black hole. They are arranged in just the same way as their namesakes near the first black hole. However, in this second case, the black hole has twice the mass of the black hole in your first experiment. The unusual clocks you had prepared as gifts for this second set of observers would need to be radically altered as for your original experiment, but the rate at which each unusual clock has to be adjusted is exactly double that of the rate needed for the corresponding clock in the first set of gift clocks at the exact same distance from the centre of the first black hole which has half the mass of the second. These time dilation effects are larger if the black hole mass is larger, and also become more extreme the closer you get to the event horizon.

Note that this time dilation is not a consequence of some additional light-travel time for a clock closer to the black hole and hence further from you, the safely-distant observer: there is not merely a time offset for an observer further away from the black hole. The closer a clock is to a black hole, the slower is the rate at which time is measured to flow on that clock, no matter what reputable means you use to measure that flow of time. Time itself is stretched (or, indeed, dilated).

What is the corollary of time dilation near a black hole? This causes effects that happen in the frame of an observer very local to the black hole to be measured to be very different from those in the frame of an observer who is very distant, worlds apart in fact.

Let's now consider what happens if in your first experiment, observer A became a little careless and dropped his first clock (the one with which he could measure proper time at his location) so that it fell towards the black hole. Despite this disaster, he would be nonetheless safely gripping onto the gift clock with which you had enticed him to participate in the experiment. Both you and A would see his first clock move towards the hole. The clock would find itself moving into the black hole, more and more rapidly. You and A would gradually notice that the time you read on the plummeting clock becomes even more discrepant with the time on A's other clock (namely the clock that was adjusted to run faster than the local clock in order that it would read the same time as the one corresponding to your time). After a while both you and A would begin to notice that time stops for the plummeting clock. A photon emitted at the event horizon towards a distant observer appears to stay there indefinitely. What happens to anything that falls into a black hole after it has passed within the critical radius of the event horizon is unknowable to an external observer. So the event horizon may be regarded as a hole in spacetime. No light will emerge from within the event horizon, as we saw in Chapter 1. That is why it is black. However, in the reference frame of the dropped clock plummeting through the event horizon, life is very far from unchanging. From the clock's perspective, it would travel to the singularity in a mere one ten-thousandth of a second, assuming that the black hole had a mass of ten times that of our Sun. If the clock had the misfortune to fall into a supermassive black hole with a mass one billion times that of our Sun (such as we meet when we study quasars in Chapter 8), its journey time inwards between the vastly larger event horizon and the singularity would be a more leisurely few hours.

Tidal forces near a black hole

Suppose in a weak moment, person A wonders about jumping, feet first towards the black hole, in hopes of being reunited with the clock he dropped. What would happen? Such a leap would prove to be a big mistake, as the survival outcome would be zero. The difference between the gravitational force on his feet and the force on his head would become extreme. This is a feature of any inverse-square force field, such as gravity from a massive body. The Earth is rather a long way from the moon, yet even the small differences in gravitational force due to the moon experienced on opposite sides of the Earth, known as tidal forces, are at the root of why the tides come and go about twice per day. In general, these forces resulting from differences in gravity in different places are called tidal forces. There are additional factors that enrich the details of the rising and falling of tides such as the gravitational force due to the relative angle of the moon, and the detailed shapes of continental masses. But even if the surface of Earth were entirely covered by ocean without land, there would still be tides with the amplitude of the sea level varying by about 20 cm twice per day, simply because of the differential gravitational force experienced by points on the planet at different distances from the Sun.

Let's now consider the smaller distance between me and the centre of the Earth. As I sit typing this chapter, my head is somewhat over a metre higher than my feet which are on the floor of my study. My feet are thus closer to the centre of the Earth than my head is. Because the gravitational force follows an inverse-square law behaviour as though all the mass of the Earth were located at the very centre of the Earth, and because my feet have a smaller distance to this centre they feel a stronger force, or pull, to the centre of the Earth than my head does. But actually, the difference is rather slender: for a height difference of one metre the difference in gravitational force is three parts in ten million. This is such a slight difference because I am about

6,400 km from the centre of the Earth. Much closer to a point mass such as a black hole, the difference in gravitational force experienced at points just a metre apart in the direction towards the black hole would be vastly more extreme. So extreme that close to the singularity A's feet would be stretched away from his knees and the rest of his body beyond what his tendons and muscles could hold together, and he would be elongated into something resembling long spaghetti. Best not to jump.

Dynamic spacetime

The rotation of a black hole makes an important difference regarding how close matter can orbit around it, and this relates to how much energy can be extracted from it. From the work of Roy Kerr and his solution to the Einstein field equations, we know that the smallest orbit that a particle can have around a black hole without falling in depends on just how fast the hole is spinning. The faster a black hole is spinning, the closer the matter can get before the hole swallows it, as illustrated in Figure 13. If you drop something straight down into a spinning black hole, it will start orbiting the hole even though there is nothing but empty spacetime outside the hole. Outside the ergosphere, it is possible

NON-SPINNING BLACK HOLE   SPINNING BLACK HOLE

13. Gas can orbit closer to a spinning black hole than to a non-rotating one.

to overcome this frame dragging using rockets, but not inside it. In the region inside the rotating black hole's ergosphere, just outside its event horizon, nothing can stand still. The spinning hole actually drags the spacetime and hence its contents around with it. A further aspect of this frame-dragging is that even if light itself is going against the direction the black hole is rotating, it will be carried in the reverse direction around the hole.

Orbiting around a black hole

It is interesting to ponder what would be the sequence of events if our Sun were to spontaneously metamorphose into a black hole right now. The first that you or I could know about it would be eight minutes later; the beautiful Spring sunlight by which I am writing would come to an abrupt halt. Although the luminosity of the single star we call our Sun is tiny by comparison with the quasars and microquasars discussed in Chapter 8, it is sufficiently close to the Earth that it provides on average about a kilowatt per square metre of power to our planet. Remarkably, this has been enough to sustain all life on the planet, allowing plants to grow and then be eaten by animals that are then eaten by other animals. The Sun has been the engine behind it all. But if fusion ceased in the Sun and it were (contrary to all expectation) to collapse into a black hole, then it would go very dark and we would all eventually die. (This is a bit of a gloomy outlook, but I encourage the reader to hold fast until Chapter 7, where we learn that our Sun is not the kind of star to form a black hole—it's too lightweight for that.) However, dynamically speaking, as far as planet Earth and the whole Solar System of planets, dwarf planets, and asteroids are concerned, nothing will change at all. All massive bodies in orbit around the Sun will continue in pretty much the same orbits. The way that gravity works is that whether the Sun has the same extent that it has now, or whether it collapses to a singularity within an event horizon of 3 km, the gravitational attraction outside the Sun would remain unchanged. The spherical collapse under gravity to a black hole would not change the angular momentum of the orbiting bodies at all, so the patterns and progressions and tides within the Solar System would continue utterly unaltered by the lack of sunshine.

Some new orbits would be possible however, much closer to the black-hole Sun than were possible previously when the solar plasma was in the way. However, these orbits could not get too close to the event horizon. The details of the warping of spacetime by a mass singularity mean that it is not possible to orbit just outside the event horizon itself. Attempting a circular orbit there would require corrective action by rockets in order to maintain the orbit. In fact, the mathematics shows that the closest that we or any other mass particle could exist on a stable circular orbit near a stationary black hole would be at a distance three times that of the Schwarzschild radius away. You have been warned.

Actually, unstable circular orbits are possible up to half this distance away from a Schwarzschild (non-spinning) black hole. This distance defines a spherical surface that is sometimes called the *photon sphere*. Even for a photon, these orbits are unstable, and before too long an orbiting photon would either slither in towards the black hole, never to return, or indeed away into space.

For a Kerr black hole though, one that has spin, the situation is different for the orbits near the black hole. In particular, there are two photon spheres, in contrast with the one photon sphere around a stationary Schwarzschild black hole. The outermost sphere is for photons that are orbiting oppositely to the direction of rotation of the black hole (the ones we say are on *retrograde* orbits). Inside this is the photon sphere for photons travelling in the same sense around the black hole as it is rotating (on *prograde* orbits). For a very slowly rotating black hole that isn't so very different from a Schwarzschild black hole, these two photon spheres are very nearly co-spatial. For black holes of increasing spin, these surfaces are increasingly further apart.

Moving closer in towards a rotating black hole, there is another important surface (discussed in Chapter 3), called the *static limit*. This is the surface at which nothing can remain static with respect to a distant observer: it is just impossible to sit still this close to a rotating black hole, no matter how powerful the rockets you might be equipped with. At this surface, even retrograde light rays are dragged along in the direction of rotation. It is still possible to escape from this close to a rotating black hole, with sufficient propulsion, but it's just not possible for anything to remain stationary and non-rotating here. Moving inwards still further, the next surface of significance is the event horizon we met in Chapter 1, the one-way membrane that we met originally in the context of Schwarzschild black holes. Crossing this outwards isn't possible and crossing it inwards has an ineluctable destiny, just as for the static black hole.

An orbit around a Kerr black hole is not generally confined to a plane. The only orbits confined to a plane are those in the plane that contains the equator (i.e. the plane of mirror symmetry of the spinning black hole). Orbits out of this equatorial plane move in three dimensions. These orbits are confined to a volume that is limited by a maximum and minimum radius and by a maximum angle away from the equatorial plane.

The details of the spin of a black hole have a dramatic effect on how close particles may encounter the black hole, which itself depends on their direction of travel relative to the spin. For a maximally spinning black hole, the photon sphere for light rays orbiting in the same sense (prograde) as the black hole spin has a radius that is half of what the Schwarzschild radius would be. For light rays on retrograde orbits, the radius of their photon sphere is twice the Schwarzschild radius. For particles with mass that are on prograde orbits, the innermost stable circular orbit on which they can move is again at half of the Schwarzschild radius. For those on retrograde orbits, such a close distance would be unstable: their innermost stable circular orbit is at 4.5 times the Schwarzschild radius. Thus, a rotating black hole enables particles on prograde orbits to orbit more closely without reaching the point of no return at the event horizon, more closely than if the black hole were non-rotating. In Chapter 7, we consider the importance of just how close matter can orbit before falling onto a black hole and how much energy may be consequently leveraged.

Chapter 5

Entropy and thermodynamics of black holes

You are what you eat

It is often said that you are what you eat. Thus if your diet is purely junk food and chocolate, then your complexion, not to mention your physical and mental well-being, will be rather different than if you subsist on a healthy diet of salad and Mediterranean food. However, it seems that black holes are not fussy eaters. Whether they are hoovering up a vast expanse of interstellar dust or a cubic light-year of fried eggs, their mass will similarly increase inexorably. In fact, after a black hole has finished its sumptuous meal, you have no way of telling what it was eating, only how much it has consumed (although you could tell if what it ate had charge or angular momentum). You only know the quantity of its diet, not about the quality. The 'no-hair theorem' described in Chapter 2 says that the black hole is only characterized by a very few parameters (mass, charge, and angular momentum), and thus we cannot talk about what the black hole is made of.

This lack of knowledge about the nature of what has been sucked in by a black hole may seem like a trivial observation, but it is actually rather profound. Information about a black hole's lunch menu has been fundamentally lost. Any matter which has fallen into the black hole has surrendered its identity. We can't perform measurements on that matter, or discern any details about it.

Black holes and engines

This situation is eerily familiar to those who have studied the beautiful subject of thermodynamics. In that field it is quite common to understand how information can become lost or dissipated through physical processes. Thermodynamics has a long and interesting history. The modern theory began during the industrial revolution when people were trying to work out how to make steam engines more efficient. 'Energy' could be defined in such a way that it was always conserved and could be converted between different forms. This is known as the first law of thermodynamics. However, although you can make some conversions between different types of energy, there are particular conversions you are not permitted to make. For example, although you are allowed to convert mechanical work completely into heat (you do that every time you use the brakes to bring your car to a complete stop), you cannot convert heat completely into mechanical work, which unfortunately is precisely what we would like to do with a steam engine. Therefore a steam engine in a train only succeeds in making a partial conversion of heat from the furnace into mechanical work which turns the wheels. It was ultimately realized that heat is a type of energy involving the random motion of atoms, while mechanical work involves the coordinated motion of some large bit of matter, like a wheel or a piston. Therefore, a crucial component of the nature of heat is randomness: because of the jiggling of atoms in a hot body, you lose track of the motion of the individual atoms. This random motion cannot simply be unrandomized without additional cost. The randomness, or to give it the technical name, *entropy*, in any isolated system never decreases but must always either stay the same or increase in every physical process. (This is the second law of thermodynamics.) One way of looking at this is to say that our information about the world always decreases because we cannot keep track of the motion of all the atoms in a large system. As energy moves from macroscopic scales to microscopic scales, from a simple moving piston to the random motion of huge numbers of atoms, then information is lost to us. Thermodynamics allows us to make this vague-sounding notion completely quantitative. This information loss turns out to be exactly analogous to what we've been describing for matter falling into a black hole.

Although thermodynamics was developed for steam engines, the principles are thought to apply to all processes in the Universe. One of the first people to think about this in connection with black holes was the Oxford physicist Roger Penrose. He reasoned that because a black hole has spin, it might be possible to extract energy from it and thus to use it as some kind of engine. He came up with an ingenious scheme in which matter is thrown towards a spinning black hole in such a way that some of it emerges with more energy than was thrown in. Energy is extracted from the region just outside the event horizon (in fact from the ergosphere discussed in Chapter 3). Penrose's process slows the rotation of the black hole. In principle, an enormous amount of energy can be extracted from a black hole in this way, but of course this is just a thought experiment and so doesn't seem to be at present a practical solution to planet Earth's looming energy crisis! Within a few years of Penrose's work, James Bardeen, Brandon Carter, and Stephen Hawking made a landmark advance and formulated what they called the three laws of black hole dynamics which laid the foundations for Hawking's later thinking on the thermodynamics of black holes, which required the concept of temperature for a black hole which is determined by its mass and spin.

Black holes and entropy

Penrose's insight was a significant impetus and got others thinking about the thermodynamics of black holes. Together with R. M. Floyd, he showed that in his imagined process the area of the black hole's event horizon would tend to increase. Stephen Hawking started working on Penrose's clever scheme. The area depends on the mass and spin (and charge) in a rather complicated way, but Hawking was able to prove that in any physical process this *area* always increases or remains the same. One of the consequences of this intriguing result is that if two black holes coalesce then the area of the black hole event horizon of the merged black holes is larger than the sum of the areas of the two original black hole event horizons. (This is intuitively reassuring because the radius of the event horizon scales with mass, and surface area has a well-known dependence on radius.) This is the same sort of behaviour that we see with entropy in thermodynamics and therefore people began to wonder whether the entropy of a black hole and its area were somehow connected. Is this more than just an interesting analogy? One of John Wheeler's students, Jacob Bekenstein, went ahead and proposed a direct connection in his PhD thesis. Bekenstein used the ideas from the information theory of thermodynamics to argue that the area of a black hole event horizon is proportional to its entropy. (The choice he made means that you take the area of the event horizon and divide by one of physicists' fundamental constants, the Planck area, which is roughly 10^{-70} square metres, and within a numerical factor you get the entropy. This choice of units makes the entropy of a black hole absolutely enormous.)

Initially Hawking didn't believe Bekenstein's results, but on further examination he was able not only to confirm the approach but deepen our understanding of how black hole thermodynamics works. It is perhaps worth understanding how these analyses are done so one can appreciate both their power but also their limitations. The ideal way forward in this field would be to use a combination of quantum mechanics and general relativity, called quantum gravity, to study systems which are both very small, like a singularity in a black hole, but in which gravity plays a big role. Unfortunately we do not have a good theory of quantum gravity at present. A good approach is to use general relativity to model how spacetime curves and then use this together with quantum mechanics to understand the behaviour of particles in the curved spacetime. This was the approach that Hawking took to attempt to understand the thermodynamics of black holes.

Is empty space empty?

The concept of the vacuum (a region where there is 'nothing' there) has had a long and tortuous history. Most of the ancient Greek philosophers hated the idea, on grounds that today seem extraordinarily arcane, but there were a small band of atomists who included the vacuum in their description of the world. Until the scientific renaissance, the idea of the vacuum was therefore very much out of fashion. However, following the invention of the air pump in 1650, the vacuum was something that you could experimentally demonstrate. Even though the amount of air that you could pump out of a vessel in the seventeenth century still gave you a rather poor vacuum by modern standards, the idea of nothingness had become substantially more believable. Once the existence of atoms had been demonstrated beyond all reasonable doubt in the early 20th century, the idea of a region of space with no atoms in it became not only uncontroversial, but inevitable.

No sooner had atoms been demonstrated than a new theory of physics arose: quantum mechanics. One of the surprising consequences of this new theory was that there were fleeting moments when it seems like energy needn't be conserved. The first law of thermodynamics, the grand and seemingly unbreakable principle of physics, insisted that at every moment and at every place there had to be a strict accountancy between energy debits and energy credits. 'Energy must always balance!' thunders the Cosmic Accountant. In fact, it seems that the Universal accountancy rules are more lenient and it is possible to obtain credit. It is perfectly acceptable to borrow energy for a short period of time as long as you pay it back quickly afterwards. The amount you can borrow depends on the duration of the loan, by an amount described by the Heisenberg Uncertainty principle. For example, even in the supposedly-empty vacuum it is possible to borrow enough energy to make a particle and anti-particle pair. These two objects can wink into existence and then after an extremely short period annihilate each other, thereby paying the energy back within the maximum allowed time limit (a time interval which is shorter the more energy is borrowed). Such a process goes on everywhere, all the time. It can even be measured! We now understand that the vacuum is actually not empty, but is a soup of these pairs of so-called virtual particles winking in and out of existence. Thus, the vacuum is not sterile and unoccupied, but is teeming with quantum activity.

Black hole evaporation and Hawking radiation

Hawking used the modern theory of the vacuum, quantum field theory, to study its behaviour close to the event horizon of a black hole. His analysis was mathematical but we can picture it in quite a simple way. The essence is that a pair of 'virtual' particles, a particle and its antiparticle (opposite in charge, identical in mass), created close to the event horizon of a black hole may end up becoming torn apart from one another. If one of that pair, either the particle or the anti-particle, falls into the event horizon it will plunge into the singularity and can be never recovered. However, its partner may remain outside the black hole. This particle has lost its virtual partner but it is now nonetheless a real particle and has the possibility of escape. If the particle does escape, rather than falling back in, it forms part of something called *Hawking radiation*. As far as a distant observer is concerned, the black hole has lost mass because a particle has been emitted. What had been realized is that, taking account of quantum field theory, black holes are not completely black, but they can actually emit particles. This argument also applies to photons, and so very weak light (also known as electromagnetic radiation) emerges from a black hole if Hawking's argument is correct.

All bodies at non-zero temperature emit thermal radiation as photons. You do this yourself, which is why you would show up on an infra-red camera even in the dark (and this is why the police and the military use such cameras). The hotter the body, the higher the frequency of the radiation. We emit infra-red radiation, but a red-hot poker is hot enough to emit visible light. Because a black hole emits Hawking radiation, it has a temperature (known as the Hawking temperature) as we have seen earlier, although this is normally incredibly low. A black hole with a mass of one hundred times that of the Sun has a Hawking temperature less than a billionth of a degree above absolute zero (which is 273 degrees below the freezing point of water)! This is one reason why Hawking radiation has not yet been detected: it is incredibly weak. But it is believed to be there.

Hawking radiation does however have an interesting consequence on the evolution of black holes: it is ultimately responsible for a black hole's eventual death. Think again about the two virtual particles. The energy of the real particle which escapes from the black hole has to be positive, but since the virtual particle pair appeared spontaneously from the vacuum, then the virtual particle sucked into the black hole must have negative energy to compensate. Because energy and mass are connected, the net effect of this process is that the black hole has had negative mass added to it, and therefore its mass will have decreased due to the emission of Hawking radiation.

Hawking had therefore discovered a mechanism by which a black hole can evaporate. Slowly, over time, the black hole will emit radiation and lose mass. This process is initially incredibly slow. It turns out that the larger a black hole is the smaller is its *'surface gravity'*. This is because even though the surface gravity depends on mass, which is larger for a big black hole, gravitational attraction follows an inverse-square law and more massive black holes are larger. The net result is that large black holes have very little surface gravity and this equates to a very low temperature. A large black hole therefore emits less Hawking radiation than a small black hole.

However, as a black hole evaporates and loses mass, the amount of Hawking radiation goes up as the surface gravity and hence temperature increases. Assuming the black hole isn't receiving any other energy, this makes the rate of mass loss faster and faster until, at the end of its life, the black hole simply pops out of existence. Thus the life of a black hole ends not with a bang but with that quiet pop. This evaporation process is only possible for black holes whose temperatures are higher than their surroundings. At the current epoch in cosmic history, the temperature of the Universe, measured from the spectral shape of the Cosmic Microwave Background radiation, is 2.7 degrees above absolute zero. Black holes with masses greater than a hundred million million kilos will not evaporate at the current epoch because their temperatures are lower than that of their surroundings. These black holes which have a slender fraction of the mass of the Sun, however, will be able to evaporate when the Universe has cooled more following further expansion. Up to this point in cosmic time, all black holes whose masses were less than one per cent of this slender value would have evaporated away by now.

The black hole information paradox

One question which arises from all of this is what happens to the information stored in the matter that fell into the black hole? One school of thought holds that this information is lost for ever, even if the black hole subsequently evaporates. Another point of view claims that that information is not lost. Because black holes evaporate, the argument goes, the information contained within the original matter that fell into the black hole must somehow be stored in the radiation from the black hole. Thus if you could analyse all the Hawking radiation from a black hole and understand it completely you would be able to reconstruct the details of all the matter that had originally fallen into the black hole. There was a famous bet between, on the one hand Stephen Hawking and Kip Thorne, and John Preskill on the other, about this very matter. Thorne and Hawking took the former position, while Preskill took the latter. The agreement was that the loser would reward the winner with an encyclopaedia of the winner's choice. In 2004, Hawking was sufficiently persuaded by the idea that information could indeed be encoded in the radiation from a black hole that he conceded the bet, supplying Preskill with an encyclopaedia about baseball (whether that constitutes a repository of meaningful information depends on your opinion of baseball); however, the matter is still debated.

Despite all these ingenious theoretical speculations, it is worth saying again that even ordinary Hawking radiation from a black hole has not yet been observed. The history of physics is littered with the relics of old, ingenious but ultimately wrong, theories. Experiments and observation have frequently been surprisingly effective at bringing forth unexpected results. Indeed, observations of spectacular phenomena have emerged that probably no-one at all would have predicted from first principles for black holes. One of the reasons that the faint Hawking radiation has not been observed is that many black holes we know about are at the centres of some of the brightest objects in the Universe, and these black holes are way too massive, and hence way too cold, to evaporate via Hawking radiation. These objects are extraordinarily bright for a completely different reason, which is examined in Chapter 6 and in Chapter 8.

Chapter 6
How do you weigh
a black hole?

The Sun, the planets that orbit around it, together with dwarf planets (of which Pluto is the most famous example), asteroids, and comets collectively comprise the Solar System. The Solar System itself orbits within the disc of our Galaxy around its centre of mass at the *Galactic Centre*. The speed at which our Solar System travels around its circular path through the Galactic disc is about 7 km/s, and to complete an entire circuit around the Galactic Centre will take a couple of hundred million years. In addition to this orbital motion, the whole Solar System moves perpendicular to the *Galactic plane*. The kind of motion it exhibits is well known to physicists as simple harmonic motion with the restoring force, which pulls our Solar System back towards the equilibrium position of the plane of the Galaxy, coming from the gravitational pull of the stars and gas that comprise the Galactic disc. At the moment, we are about 45 light-years above this equilibrium point. In about 21 million years from now the Solar System will be at its extreme point 320 light-years above the Galactic plane. 43 million years after that, the Solar System will be back in the mid-plane of the Galaxy. When the Solar System lies in the centre of the Galactic plane then, the Earth will suffer maximum exposure to the cosmic rays that are whizzing around in the plane of the Galaxy, trapped along lines of magnetic field, and travelling around them on some kind of a cross between a helter-skelter and a tramline. There have been speculations that the Sun's motion through the Galactic plane could have been responsible for the mass exinction of dinosaurs. But this kind of speculation is hard to verify or refute because the timescales for this orbital motion are of course rather tricky for human observers, who don't tend to live longer than one century. This is a common problem in observational astronomy when we want to follow some process that changes on timescales much longer than the few centuries over which we've been making astronomical observations of any reasonable accuracy and thoroughness.

There are, however, orbital motions within the Galaxy that are significantly easier to measure, at least in the sense that the relevant timescales are commensurate with the attention spans of humans and their telescopes. Of particular interest in the context of black holes are the orbital motions of the stars in the innermost regions of the Milky Way, that appears in a part of the sky known as Sagittarius A*. Looking into this region, most easily seen from the southern hemisphere, one is looking towards the very centre of our own Galaxy, 27,000 light-years away from us. This is a particularly densely populated region of space, which leads us to two problems when we want to study the Galactic Centre. The first is that there is a relatively high space density of stars and the second is that there is lots of dust.

The first problem means you need to use a measurement technique that enables high resolution imaging, i.e. fine details can be separated from one another in the way that a telephoto lens gives finer detail on a given camera than a wide-angle lens does. Just using a larger telescope is invariably insufficient for this, but there are various techniques developed for untangling the turbulence in Earth's atmosphere through which we inevitably view all celestial objects, unless we put the telescope on a satellite above the atmosphere. Of particular importance is a technique known as *adaptive optics*. This technique corrects for atmospheric variations by observing the blurring of a bright star (called a guide star) and deforming the primary mirror of the telescope to cancel out this varying blurring. When a bright star isn't available in the part of sky that is of interest, a powerful collimated laser beam can be shone up to excite atoms in the atmosphere and the atmospheric corrections derived from that.

The second issue, the presence of vast quantities of dust towards the Galactic Centre, is problematic because it is hard to see optical light through dust, just as it is hard for ultra-violet light from the Sun to penetrate through the opacity of a sunhat. The solution to this problem is that one needs to observe at infra-red wavelengths rather than visible wavelengths.

How to measure the mass of the black hole at the Galactic Centre

Such infra-red observations have been championed by two groups, one led by Andrea Ghez in California and one led by Reinhard Genzel in Germany. The work of both teams independently provides a wonderfully clear measurement of the mass at the centre of the Galaxy. Figure 14 shows the data from Andrea Ghez and her team. Over the last few years they have made repeated observations right into the very heart of the Galactic Centre and watched how the stars have moved since the last time they observed them. Because the spectral types of these stars are known, their masses are known. Year by year, as the orbital path of each of these stars becomes apparent, the dynamical equations (known as Kepler's laws, the same laws that govern the motion of the planets around our Sun) enable Ghez and her team to solve for each orbit independently and deduce the mass of the 'dark' region that is at the common focus of all these orbits. These independent solutions determine the mass of this dark region rather well. It is now known to be just over 4 million times the mass of our Sun, within a region whose radius is no more than 6 light-hours. Because the object is dark but extraordinarily massive, the only conclusion is that there is a mammoth black hole at the centre of our Galaxy.

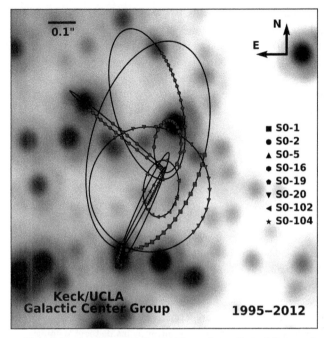

14. Figure showing the successive positions of stars that orbit around the central black hole in our Milky Way.

There is no reason to believe that our Galaxy, the Milky Way, is unique in having a black hole at its centre. On the contrary, it is strongly suspected that all galaxies may well have a black hole at their centres, at least the more massive ones. The reason for this is because of a seemingly fundamental relationship, discovered by John Magorrian, then at the University of Durham, and co-workers, between the mass of a black hole at the centre of a galaxy and the mass of the galaxy itself. Of course the business of measuring the mass of a black hole and the mass of a galaxy is tricky. The technique that works so beautifully at the centre of our Galaxy cannot be applied to external galaxies because they are simply too far away.

The masses of the central black holes at the hearts of elliptical galaxies exceed a million times the mass of our Sun and indeed extend up to and beyond a billion times the mass of our Sun. For this reason, they are often termed *supermassive black holes*.

Despite the difficulties in measuring the masses of black holes and the masses of galaxies, it has been found for a wide range of different galaxies that the mass of the central black hole scales with the mass of its host galaxy. This is thought to suggest that both the central black hole and the galaxy itself grew and evolved together across cosmic time.

Many black holes throughout the Galactic disc

Besides the single, central supermassive black hole at the heart of a galaxy, there are thought to be millions of black holes distributed throughout the extent of each galaxy, and these are believed to have formed in a very different way from the galactic-central ones which grow by gradual accretion of infalling matter. These *stellar mass black holes* are formerly massive stars, once shining very brightly, with fusion powering away inside them keeping them very hot and pressurized, and crucially able to resist gravitational collapse. When their nuclear fuel is all used up, there is no longer any radiation pressure to hold up the star, and nothing to balance the inward force of gravity. For a star with a similar mass as our Sun, the collapse under gravity ultimately results in a compact object known as a white dwarf. The word *compact* has special meaning in astrophysics and connotes that the matter is dense in a way that is utterly distinct from normal matter. By the standards of normal matter, white dwarfs are compact because the matter has been extremely compressed. This matter is ionized, meaning that all the electrons are separate from their parent nuclei, yet cold (normally matter is only ionized at high temperature). The pressure that withstands the persistent inward gravitational pull arises from the electrons refusing to be compressed into too confined a region (a consequence of the

Heisenberg uncertainty principle); the technical name for this effect is 'electron degeneracy pressure'. Had the collapsing star, when it had used up all its fuel, been more massive, then the gravitational infall would have been greater still and the electrons and their counterpart protons would have fused together to form neutrons. These can form a much more compact object than a white dwarf—a neutron star.

But, if we are interested in black holes, then we must turn to stars which are considerably more massive than those which go on to produce white dwarfs or even neutron stars. A star above this mass will be very luminous while its fuel lasts and nuclear fusion can be sustained. Once all the fuel is used up, it's game over for the star and the lights will switch off. The star is now sufficiently massive that the gravitational force can overwhelm even the strong neutron degeneracy pressure and so the collapse is so powerful that even this pressure cannot balance gravity and the collapse leads inexorably to a black hole. The collapse of a massive star is often accompanied by the explosion of a spectacular supernova remnant, leaving a black hole as the only remnant at the original location of the progenitor star. In such explosions many elements, particularly those heavier than iron, are synthesized.

The first black hole to be securely identified from a determination of the masses of the two stars in a binary star system is called V404 Cyg. Jorge Casares and Phil Charles and their co-workers observed the orbits of the two stars very carefully and inferred from their analysis that this binary pair includes a compact object having a mass at least six times greater than the mass of our Sun, and is thus a black hole. (Its mass was later found to be twelve times the mass of the Sun.)

It is possible to make plausible estimates of the numbers of stars in galaxies and their masses. We can then estimate the number of 'stellar-mass' black holes in our Galaxy by considering how many massive stars would have formed early enough in its history to have evolved sufficiently by now to use up all their nuclear fuel via fusion. Even if only a very small proportion of stars in our Galaxy go on to form black holes, with more than 10^{11} objects in the Milky Way that still gives us a lot of black holes.

How can one measure the masses of these black holes that pervade galaxies? In fact for some stellar-remnant black holes, the technique is dynamically very similar to that used for the black hole at the centre of our Galaxy. The reason for this is that a very significant fraction of stars in our Galaxy, and therefore most probably in other galaxies also, come in pairs that formed binary star systems. It is easy to surmise how this might come about: gravitational forces are attractive and many two-body orbits are stable, so once two stars encounter one another and become gravitationally bound together, they are likely to remain so. For a binary system, if we can measure the time taken for the stars to do a complete loop around one another, a time known as the orbital period, and if we know the distance between them, then we are well on the way to finding their masses. If the compact object is in orbit around a normal (fusion fuelled) star of known spectral type and therefore known mass, then the mass of the compact star is straightforward to derive. If a compact object such as a black hole is a singleton and not in a binary, then the lack of dynamical information means that there is no means of inferring its mass and or indeed of determining that it is a black hole. The smallest black hole that we can measure is a few times the mass of our Sun, but the heaviest stellar-mass black holes can exceed a hundred times the mass of our Sun.

The measurement of the mass of a black hole, given modern day technology, is tractable although it still requires a good measure of patience and tenacity. Given that mass is one of essentially only two fundamental physical properties of a black hole such studies get us half-way to characterizing it! However, measuring the spin of a black hole is harder, and in Chapter 7 I describe the heroic efforts that are needed to try and do this.

Chapter 7
Eating more and growing bigger

How fast do they eat?

The popular notion of a black hole 'sucking in everything' from its surroundings is only correct near the event horizon, and even then, only if the angular momentum of the infalling matter isn't too great. Far away from the black hole, the external gravitational field is identical to that of any other spherical body having the same mass. Therefore, a particle can orbit around a black hole in accordance with Newtonian dynamics, just as it would around any other star. What could unravel this pattern of going round and round in circles (or indeed ellipses) and pave the way for more exotic behaviour? The answer is that there is invariably more than one particle orbiting the black hole. The richness of the astrophysical phenomena we observe arises because there is a lot of matter orbiting around a black hole and this matter can interact with itself. What is more, gravity isn't the only law of physics that must be obeyed: so too must the law of conservation of angular momentum. Applying these laws to the bulk quantities of matter that may be attracted towards the black hole gives rise to remarkable observable phenomena, good examples of which are found in the case of exotic objects known as quasars. *Quasars* are objects at the centres of galaxies having a supermassive black hole at their very heart which, because of its effect on nearby matter, can cause it to outshine the collective light from all the stars in one of those galaxies, across all parts of the electromagnetic spectrum. We shall meet quasars, and other examples of 'active galaxies', in Chapter 8, together with scaled-down counterparts of these called microquasars whose black holes are orders of magnitude less massive than those inside quasars. For now let's get back to thinking about the matter around a black hole.

As we have noted, you cannot directly observe an isolated black hole because it simply won't emit light; you can only detect a black hole by its interactions with other material. Any matter falling towards a black hole gains kinetic energy and by *turbulence*, that is to say swirling against other infalling matter doing a similar thing, becomes hot. This heating ionizes the atoms leading to the emission of electromagnetic radiation. Thus, it is the interaction of the black hole on the nearby matter that leads to radiation being emitted from the vicinity of the black hole, rather than direct radiation from the black hole itself.

Black holes are not aloof, non-interacting entities in space. Their gravitational fields attract all matter, whether nearby gas or stars, towards them. Because gravitational attraction increases strongly with proximity, stars are ripped apart if they are unfortunate enough to have a close encounter with a black hole; an example is pictured in Figure 15. A certain fraction of the attracted matter will be entirely swallowed or *accreted* by the black hole. Matter doesn't just accelerate into the black hole whooshing through the event horizon. Rather, there is something of an elaborate courtship ritual as the gravitationally-attracted matter draws near the black hole. Very often it is found that a particular geometry characterizes accreting matter: that of a disc. If the gravitational field were spherically symmetric, the black hole would play no role in determining the plane within which the gas would settle to form an accretion disc—the disc plane would be determined by the nature of the gas flow far from the black hole. If, however, the black hole has spin, accreted gas will eventually settle into the plane perpendicular to its spin axis, regardless of how it flows at

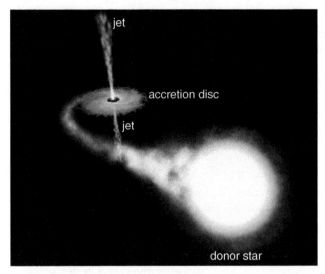

jet

accretion disc

jet

donor star

15. Artist's impression of an accretion disc (from which a jet is shown to emanate—see Chapter 8) and a donor star which is being ripped apart by the gravitational tidal forces from the black hole which is at the centre of the accretion disc.

large radii. If there is any rotation at all in the attracted matter, then this must be thought of in terms of the conservation of angular momentum that we met in Chapter 3 when we considered the rotation of material that ultimately collapsed to form a spinning black hole. The rotation means that the matter will be following (fairly circular but actually) spiralling-in orbits as it loses energy. Close to the black hole, the Lense–Thirring effect that we met in Chapter 3 means that at small radii the accretion disc may become aligned with the equatorial plane of the spinning black hole. (In this context, this effect is known as the Bardeen–Petterson effect.)

If gas is a significant component of the collapsing matter then gas atoms can collide with other gas particles on their own orbits and these collisions result in electrons in those atoms being excited to higher energy states. When these electrons fall back to lower energy states they release photons whose energies are precisely the difference between the higher energy level of the electron and the lower energy level it has fallen to. The release of photons means that radiative energy leaves the collapsing gas cloud and so this loses energy. While energy is released in these processes, bulk angular momentum is not. Because angular momentum remains in the system, the coalescing matter continues to rotate in whatever plane conserves the direction of the original net angular momentum. Thus, the attracted matter will invariably form an accretion disc: a rather long-lived holding pattern for material orbiting the black hole. Depending on just how close to the black hole the orbiting material can get, the matter can get so hot that the radiation emitted from the accretion disc actually comprises X-ray photons, corresponding to high temperatures of ten million degrees (it doesn't matter too much whether the Kelvin or Celsius temperature scale is being used when the temperatures are quite this hot!).

A simple analysis of some familiar equations from Newtonian physics shows that the gravitational energy release for a given amount of infalling mass depends on the ratio of its mass multiplied by that of the black hole it is spiralling towards, and how close to the black hole the infalling mass gets. For a given mass of attractor such as a black hole, the closer the infalling mass approaches it, the greater the gravitational potential energy released as can be seen in the cartoon in Figure 16. The energy that is available to be radiated out is the difference between the energy the infalling mass has far away before it is accelerated (calculated using Einstein's famous formula $E = mc^2$, where E is energy, m is mass, and c is the speed of light) and the energy it has at the innermost stable circular orbit of the black hole.

Although fusion holds great hope as a future source of energy for Earth, it can only yield at most 0.7% of the available '$E = mc^2$' energy. In contrast, significantly more of the available rest mass can be released as energy from accreting material, via

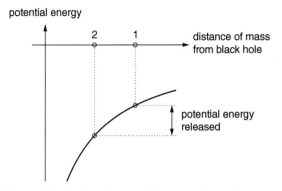

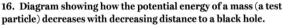

16. Diagram showing how the potential energy of a mass (a test particle) decreases with decreasing distance to a black hole.

electromagnetic or other radiation. Quite how close to a black hole the accreting material can get depends, as described in Chapter 4, on how fast the black hole is spinning. If the black hole is spinning fast, the holding pattern of the material can be orbiting much closer in, on much smaller orbits. In fact, accretion of mass onto a spinning black hole is the most efficient way known of using mass to get energy. This process is thought to be the mechanism by which quasars are fuelled. Quasars are the sites of the most powerful sustained energy release in the Universe and are discussed further in Chapter 8.

I've already mentioned there is an equivalence between mass and energy and for a Schwarzschild (non-rotating) black hole, an amount of energy equivalent to 6% of its original mass could in principle be liberated, and that Roy Kerr's solutions to the Einstein field equations show that the last stable circular orbit has a much smaller radius from the spinning black hole than would a non-rotating black hole of the same mass. In principle, vastly more rotational energy can be extracted from a Kerr black hole, but only if the infalling matter is orbiting in the same sense as the black hole itself. If matter is orbiting in the opposite direction to the way the black hole is spinning, i.e. it is on a retrograde orbit, then not quite 4% of the rest energy could be released as electromagnetic radiation. If, however, the matter infalling towards a maximally spinning black hole were orbiting in the same sense as the black hole were spinning, then in principle a remarkable 42% of the rest energy could be released as radiation, if the matter could lose sufficient angular momentum that it could orbit the black hole as close as the innermost stable prograde circular orbit.

How fast do they eat?

The accretion rate of the black hole at the centre of our Galaxy, in Sagittarius A*, whose discovery we met in Chapter 6, is 100-millionth of the mass of the Sun per year. This doesn't sound very much until you realize that this corresponds to an appetite of 300 Earth masses per year. To account for the typical, immense luminosities of quasars, matter-infall rates amounting to a few times the mass of our Sun each year are required. To account for the typical luminosities of the smaller-scale microquasars that we shall also meet in Chapter 8, the required matter-infall rates might be one millionth of this value.

Another context in which a similar energy extraction process may be taking place is in *gamma-ray bursts*, usually referred to as GRBs. These are sudden flashes of intense beams of gamma rays that seem to be associated with violent explosions in distant galaxies. They were first observed by US satellites in the late 1960s and the received signals were initially suspected to be from Soviet nuclear weapons.

Given the ubiquity of matter spiralling into a black hole via a disc, physicists find it helpful to make simple and instructive calculations to get a handle on the magnitudes of some of the important physical quantities: if' one considers a spherical geometry rather than a disc geometry then some interesting limits emerge. A particularly illustrative example comes from the world of stars, which are much better approximations to spheres of plasma than are accretion discs. Sir Arthur Eddington pointed out that the radiation released by the excited electrons colliding with other ions in the hot gas of a star will exert a radiation pressure on any matter that it subsequently intercepts. Photons can 'scatter' (which simply means 'give energy and momentum to') electrons contained in the hot ionized plasma within the interior of a star. This outward pressure is communicated via electrostatic forces (the electrically-charged analogue of the gravitational force) to the positively charged ions such as the nuclei of hydrogen (also known as protons) and the nuclei of helium and other heavier elements that are present.

In the case of a star, the net radiation heads radially outwards and this resulting outward radiation pressure acts oppositely to the gravitational force that pulls matter inward towards the centre. For the more-or-less spherical geometry of a star, there is a maximum limit to the amount of outward radiation pressure before it overwhelms the inward gravitational pull and the star simply blows itself apart. This maximum radiation pressure is known as the *Eddington limit*. Higher radiation pressure inevitably follows from higher luminosity of radiation, and the luminosity of an object can be estimated from its brightness if we know the distance to the object. Therefore, with certain simplifying assumptions including approximating an accretion disc to a sphere, the amount of radiation pressure inside an object can be inferred. This simple method is sometimes used to make an indicative estimate of the mass of the black hole: from the observed luminosity of the radiation to emerge from the surrounding plasma, if it is deemed to be at the maximal limiting value of the 'Eddington luminosity' (above which higher luminosity would give sufficiently high radiation pressure that it would exceed the gravity from the mass within and hence blow itself apart) then the mass can be estimated.

This Eddington luminosity can be thought of in terms of a maximum rate at which matter can accrete, for suitable assumptions about how efficient the process of accretion is. This gives a quantity called the Eddington rate which (for the assumed efficiency) is a maximal value. There are ways of breaking this particular maximum limit, not the least of which is the rejection of the assumption of spherical symmetry (this is fine for a star but manifestly doesn't apply to the disc-geometry of accretion discs that we need to consider in order to understand how black holes grow).

How to measure the speed of rotation within an accretion disc

Because of advances in astronomical technology it is now possible to measure the speed at which material is orbiting a black hole, at least for examples that are relatively close to Earth. One of the big challenges is that it is difficult to obtain information on a sufficiently fine angular scale. The spatial resolution required needs to be at least one hundred, if not one thousand, times finer than that routinely obtained by optical telescopes. In principle, the route to achieving finer resolution with a telescope would be to observe at shorter wavelengths and to build a larger telescope, in particular to reduce the ratio of the wavelength of observation to the diameter of the telescope being used. Unfortunately, the latter gets hideously expensive very quickly while the former takes the usual visible observing wavelengths into the ultra-violet regime, to which the atmosphere of the Earth is rather opaque. The route to achieving a smaller ratio of observing wavelength to telescope diameter is, counter-intuitively, to observe at radio wavelengths (much longer wavelengths than either visible or ultra-violet) for which the atmosphere and ionosphere are usually transparent, and to take the telescope diameter to be most of the Earth's diameter.

There are a few technical issues about this approach which need a little discussion: it turns out that thanks to some very useful mathematics developed by the French mathematician Jean Baptiste Joseph Fourier, it is possible to recover much of the signal that a full telescope aperture would observe, even if the actual

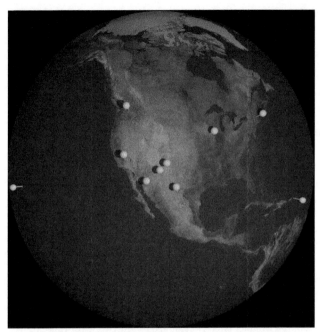

17. Artist's impression of the Very Long Baseline Array (VLBA) of antennas that collectively give images with a resolution equal to that which would be obtained by a telescope with an aperture a significant fraction of the diameter of the Earth.

collecting area only exists in a sparse subset of the full aperture that one would ideally prefer. If the signals from discrete antennas (each looking like an individual telescope—see Figure 17 showing the Very Long Baseline Array, known as the VLBA) are correlated together, it is possible to reconstruct images of small regions of the sky that have detail as fine as that which would be obtained if an Earth-sized telescope could have been fully built. Just to give an idea of how fine this resolution is, suppose that I was standing on top of the Empire State Building in New York, and you were in San Francisco. With this amount of resolution you would be able to resolve detail that is separated by the size of my little finger nail.

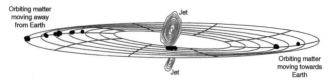

18. The VLBA has measured the distribution of discrete masers orbiting within the accretion disc of the galaxy NGC 4258 (also known as Messier 106) around its central black hole whose mass is 40 million times the mass of our Sun.

(I am glossing over the fact that the Earth is a sphere so there is no direct line of sight between San Francisco and the Empire State Building, but you get the idea.) This means that with instruments like the VLBA we can see individual features less than a light-month apart in other galaxies.

High resolution across an image in a spatial sense, and high resolution in a spectral sense (meaning that one can discern very precisely what the wavelengths of particular features are in a spectrum) is a very powerful combination. Making use of the Doppler effect, a team led by Jim Moran of Harvard University used the VLBA to make observations of the accretion disc surrounding the central black hole of a nearby galaxy known as NGC 4258. They measured the variation in wavelength of a particular spectroscopic signal (a *'water maser'*) across the rotating accretion disc and used this redshifting and blueshifting, as the masing matter moved towards and away from the Earth, to detect the variation in the speed with which matter at a given distance orbits around the black hole. These exquisitely beautiful data confirm that the matter orbits around the black hole just as Kepler's laws would describe, and these orbits are depicted in Figure 18.

Swirling matter

In the innermost stable orbit of a black hole whose mass is 100 million times the mass of our Sun, the angular momentum is over 10,000 times smaller than the angular momentum of matter orbiting in a typical galaxy. It is clear that for matter to be accreted by the black hole, this requires the removal of the vast majority of this angular momentum, and this is accomplished by processes within the accretion disc. The orbits in an accretion disc may be regarded as a good approximation to circular although in fact they are subtly and gradually spiralling in. Kepler's laws say that the matter orbiting on the smaller radii will be moving faster than the matter on slightly larger orbits. This differential rotation allows a black hole to absorb the plasma that comprises the accretion disc: the rapidly rotating inner orbits friction burn against the neighbouring material on orbits with slightly larger radii. This difference in velocity will mean that the matter on slightly larger orbits will, by viscous turbulence effects, be dragged along a little faster and so correspondingly the matter on inner orbits will be slightly slowed. Therefore, because orbital motion has increased further out, angular momentum has been transferred to the outer material from the inner material, heating as it does so.

Overall, angular momentum is conserved, and the inner material can systematically lose angular momentum, making it more likely to be swallowed by the black hole. Note that if a blob of orbiting matter has too much angular momentum, it will stay further away from the centre of mass about which it is orbiting: it would be moving too fast to get any closer. What kind of viscous effects might be relevant to the plasma within an accretion disc? Inter-atomic viscosity can be small in this situation—the gaseous plasma of which the accretion disc is comprised is very far removed from the consistency of treacle. In fact, magnetic fields may be very important in transferring angular momentum out of accreting inflow. Where do the magnetic fields come from? The plasma in an accretion disc is very hot, and so the atoms are partially ionized into electrons and positively charged nucleons. Therefore, there are flows of charged particles and moving charges produce magnetic fields, as described by the equations of James Clerk Maxwell. Once even very weak magnetic fields exist, they can be stretched and amplified by differential rotation and modified by the turbulence of the plasma, up to levels at which they can give the required viscosity. This is the basis of what is known as the magnetorotational instability. The importance of this mechanism in this context was first realized by Steve Balbus and John Hawley in the early 1990s when working at the University of Virginia.

By viscous turbulence and probably other means, plasma can eventually lose angular momentum and orbit at smaller radii closer to the black hole. Once the gaseous plasma reaches the innermost stable orbit, no more friction is needed for it to slip down into the black hole, after which it will never be seen again, but it will have augmented the mass and spin of the black hole.

What do accretion discs look like, and how hot are they?

We have seen that viscous and turbulence effects play a significant role in removing angular momentum from the orbiting material so that it can orbit more closely to the black hole and be swallowed by it. A consequence of the viscous action, however, is that the bulk orbital spiralling motion gets converted into random thermal motion and hence the matter heats up. The greater the random thermal motion of matter, the more heat energy it has and the higher its temperature. As mentioned in Chapter 5, wherever there is heat, there will be thermal electromagnetic radiation. Every body emits thermal radiation, unless it is at absolute zero.

Such heating is what is responsible for the highly luminous radiation we observe from accretion discs. For the accretion discs that surround the supermassive black holes that are at the hearts of quasars, the characteristic size of an accretion disc is a billion kilometres and the bulk of the radiation from these accretion discs is in the optical and the ultra-violet region of the spectrum. For the accretion discs that surround the vastly less massive black holes in

the so-called microquasars (that are discussed in Chapter 8),
the accretion discs are a million times smaller in extent and the
radiation is dominated by X-rays. The more massive a black hole
is, the larger the innermost stable circular orbit is and hence the
cooler the surrounding accretion disc will be.

The maximum temperature in an accretion disc around a
supermassive black hole 100 times the mass of our Sun will be
around 1 million Kelvin while for a disc around a stellar-mass
black hole, it can be up to a factor of 100 higher.

How do you measure how fast a black hole is spinning?

Given you can't actually directly see black holes, you can't see them
spinning either. But there are nonetheless two main routes to
measuring how fast a black hole is spinning. As discussed in
Chapter 4, when black holes spin very fast, it is possible for matter
to be in stable orbit around the black hole much closer in than
would be possible if they were not spinning. It turns out that
matter in these very tight orbits is heated by strong turbulent and
viscous effects as it swirls in, and this immense heat can lead to
X-rays being emitted, depending on how close to the black hole
the matter has swirled in before being swallowed up. General
relativity predicts that the shape of the spectral lines is affected by
the distance the emitting matter is from the black hole (arising
from the gravitational redshift) in a way that has a characteristic
signature. This signature arises from fluorescing iron atoms within
this matter and the method of extracting information from X-ray
light was pioneered by Andrew Fabian of Cambridge University.

These are challenging measurements to interpret, because of
many different factors, such as the inclination of the accretion disc
with respect to Earth, and indeed the nature of wind and
outflowing matter from the surface of the accretion disc, in the
vicinity of (along our line of sight to) its inner rim whose

characteristics hold the key to unlock information about the black hole that is otherwise inaccessible. Other methods for measuring the spin of stellar mass black holes involve measuring a significant range of the X-ray spectrum and accounting for the different temperatures of the inner regions of the accretion disc (which are hotter) and the regions further out (that become gradually cooler). It is possible to estimate from the shape of the X-ray spectrum the inclination of the disc and from the highest temperature (assuming you know the mass of the black hole, and its distance from Earth) to infer how far from the black hole the innermost material is orbiting. Analogous methods to measure the spin of supermassive black holes at the hearts of quasars are being developed by Christine Done at Durham University. How close that matter is able to orbit (before being swallowed by the black hole) tells you how fast the black hole itself must be spinning.

Black holes are very messy eaters

It transpires that only a fraction (estimated to be 10%, though it can be very significantly higher) of the matter that gets attracted in towards a black hole gets as far as the event horizon and actually gets swallowed. Chapter 8 considers what happens to the matter infalling towards a black hole that doesn't actually get swallowed within the event horizon. From across the accretion disc itself, matter can blow off as a wind; from within the innermost radii of the accretion disc very rapid jets of plasma squirt out at speeds that are really quite close to the speed of light. As Chapter 8 shows, what doesn't get eaten by the black hole gets spun out and spat out rather spectacularly.

Chapter 8
Black holes and spin-offs

Black holes don't just suck

If our eyes could observe the sky at radio or at X-ray wavelengths, we would see that some galaxies are straddled by vast balloons or lobes of plasma. This plasma contains charged particles that move at speeds close to the speed of light and radiate powerfully across a range of wavelengths. The plasma lobes exhibited by some of these galaxies (examples of 'active galaxies') are created by jets, travelling at speeds so fast that they are comparable with the speed of light, that are squirted out from the immediate surroundings of a black hole, outside its event horizon. Roger Penrose showed in general terms how extraction of the spin energy of a black hole from its ergosphere might be possible in principle. Roger Blandford and Roman Znajek have shown explicitly how the energy stored in a spinning black hole could actually be transferred into electric and magnetic fields and thereby provide the power to produce these relativistic jets of plasma. There are also other explanations for the mechanism by which jets are launched from near black holes. However, which of these is correct is the subject of active and exciting current research.

Whatever the mechanism(s) turn out to be, these jets are highly focused, collimated flows ejected from the vicinity of the black hole, but of course outside the event horizon. The regions in between galaxies are not, in fact, empty space. Instead they are filled with a very diffuse and dilute gas termed the *intergalactic medium*. When the jets impinge on the intergalactic medium, shock waves form within which spectacular particle acceleration occurs, and the energized plasma which originated in a jet from near the black hole billows up and flows out of the immediate shock region. As the plasma expands, it imparts enormous quantities of energy to the intergalactic medium. There are many instances of these plasma jets extending over millions of light-years. Thus black holes have tremendous cosmic influence, many light years beyond their event horizons. In this chapter, I will describe the influence and interactions of black holes on and with their surroundings.

As discussed in Chapter 6, at the centre of (probably) most galaxies is a black hole, on to which matter accretes, giving rise to emission of electromagnetic radiation. Such galaxies are called active galaxies. In some of these galaxies, the process of accretion is extremely effective and the resulting emission of radiation extremely luminous. Such galaxies are called quasars (a term which derives from their original identification as 'quasi-stellar radio sources', vastly distant, highly luminous points of radio emission). We now understand that quasars are the sites of the most powerful sustained energy release known in the Universe. Quasars radiate energy across all of the electromagnetic spectrum, from long wavelength radio waves, through optical (visual) wavelengths, to X-rays and beyond. The radio lobes, mentioned above, can be especially dramatic because they extend across distances of over hundreds of thousands of light-years (see Figure 19). The energy radiated at radio wavelengths arises from those large lobes—reservoirs of ultra-hot magnetized plasma, powered by jets that transport energy over vast distances in space. Highly energetic electrons (highly energetic here meaning travelling extremely close to the speed of light) experience forces across their direction of travel from the ambient magnetic fields that pervade the plasma lobes within which they are travelling.

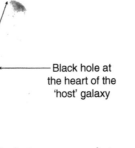

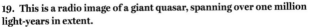

I million light-years

Black hole at
the heart of the
'host' galaxy

19. This is a radio image of a giant quasar, spanning over one million light-years in extent.

This acceleration causes them to emit photons of radiation (which may be radio, or in rare, highly energetic instances, at shorter wavelengths still, all the way up to X-rays) known as *synchrotron radiation*.

To give a sense of the scale of the power produced by quasars, consider the following values. The LEDs by whose light I am working have a power output of ten watts. They are illuminated by electricity from my local power station which produces a few billion watts (a billion watts is 10^9 watts or a gigawatt). The Sun outputs about 4×10^{26} watts, more than a hundred million billion times that from this power station. Our Galaxy, the Milky Way, contains more than a hundred billion stars, and its power output is approaching 10^{37} watts. But the power produced by a quasar can exceed even the Galactic power output by more than a factor of 100. Remember, this power is being emitted not by a galaxy of one hundred billion stars but by the processes going on around a single black hole. Such radiation could do considerable damage to the health of living creatures here on Earth, so it is just as well for us that there are no examples of such powerful quasars too near our Galaxy!

Jets in quasars are thought to persist for a billion years or less, an idea that comes from estimates of the speed at which these objects' jets grow and from measurements of the size they have grown out to. A simple relationship between distance and time and speed therefore gives a guide to the likely durations of jet activity in the quasars that are observed across the cosmos.

As these radio-emitting lobes expand, their magnetic fields weaken as do the 'internal' energies of the individual electrons in the lobes. These two effects serve to diminish the intensity of the radiation with time and with distance from the black hole; how dramatically this intensity falls off depends on how many highly energized electrons there are compared with how many less energetic ones there are. It's a property of synchrotron radiation that the lower the magnetic field strength is, the more energetic the electrons need to be to produce the radiation at the wavelength that your radio telescope is tuned to receive at. This compounds the diminishing of the synchrotron radiation as the plasma lobes expand into outer space. Not only do the electrons lose energy as the plasma expands, but because the magnetic field strength is weakening, only increasingly energetic electrons are relevant to what is observed by your telescope and, very often, there are vastly fewer of these than there are of the lower energy electrons anyway. As far as radio lobes of quasars are concerned, the lights can go out really quite rapidly.

The show isn't over, but the spectacle does move over to a different waveband. Something rather remarkable happens: the lobes light up in X-rays. This happens via a scattering process known as *inverse Compton scattering*. In the presence of a sufficiently large magnetic field, electrons can emit synchrotron radiation and thereby lose energy. Another mechanism of losing energy that is relevant to our discussion here happens via the interaction of these electrons with photons that comprise the Cosmic Microwave Background (CMB), the radiation that is left over from the Big Bang and which currently bathes the Universe in a cool microwave glow. It is possible for such an electron to collide with a photon from the CMB so that the photon ends up with a lot more energy than it had before the collision and the electron ends up with a lot less energy than it had before the collision (energy is conserved overall, remember). Of particular interest is that when the energies of the rapidly moving electrons reduce to a mere one thousand times the energy of an electron at rest (having previously been a hundred or a thousand times higher than this) their energies are perfectly matched so that they will upscatter CMB photons into the X-ray photons. The interaction of an energetic electron with a low-energy photon to yield a high-energy photon is somewhat analogous to the situation in snooker where the white cue ball (imagine this is an electron) collides with one of the red snooker balls (for the purposes of this illustration please overlook the fact that this ball isn't moving at the speed of light!) and the red ball gains a lot of energy at the expense of the cue ball. Whereas (hopefully) the red ball ends up in one of the pockets on the snooker table, the photon (which originally had a wavelength of about a millimetre) acquires about one million times as much energy as it had before the collision so that its wavelength becomes a million times shorter.

The *Chandra* satellite, launched by NASA in 1999, is sensitive to X-ray wavelengths and in fact can detect pairs of dumb-bell lobes in the X-rays just as a radio telescope can detect these double structures at cm-wavelengths. Figures 20 and 21 show in contour form double structures observed at radio wavelengths and in greyscale form the double structures in X-rays.

In fact if we were able to monitor the life cycle of one of these quasars throughout all these evolutionary stages (analogously to how a biologist might observe the life cycle of the frog from frogspawn, to tadpoles, to tadpoles with little legs, to little frogs with stumpy tails, to larger frogs to dead frogs) we would observe a cross-over from the double structures being radiant at radio wavelengths to becoming increasingly dominant in the X-ray region. First the radio structures would fade beyond detectability then the X-ray structures would fade beyond detectability. Of course, if the jets were to re-start, for example if the black hole

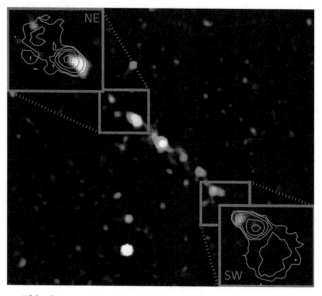

20. This giant quasar is half a million light-years in extent, and has a double-lobe structure at both radio (shown as contour lines) and X-ray (shown in greyscale) wavelengths.

were to get more fuel, then the jets would fuel new radio-emitting double lobes and then X-ray emitting lobes again. As we have seen in Figures 20 and 21, in some quasars we can see both the radio and the X-ray double structures at the same time but in others, only one or the other (Figure 22). In a couple of remarkable cases we see the X-ray double structure corresponding to a previous incarnation of jet activity, but also some new radio activity, at a different angle because the direction along which the oppositely-directed jets are launched has swung round, i.e. it has precessed; an example of this phenomenon is seen in Figure 21.

The steadiness of the jet axis of many quasars and radio galaxies is a pointer to the steadiness of the spin of the supermassive black hole, acting like a gyroscope. Why some of these jet axes should

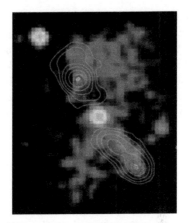

21. The double-lobe structure observed in this quasar at radio wavelengths [contours] showing the more recent activity to be differently oriented from that showing at X-ray energies [greyscale] (the relic emission revealed by inverse Compton scattering of CMB photons) suggesting that the jet axis may have precessed as the jet axes in microquasars do.

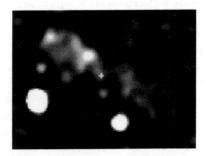

22. This is an X-ray image and shows the double-lobe structure straddling this galaxy which is only detectable at X-ray wavelengths.

precess but not others will be answered when we can discover what controls the angular momentum of the jets at the launch point near the black hole. Whether this is the spin axis of the black hole itself, or whether it is the angular momentum vector of the inner part of the accretion disc, compounded no doubt by the Lense–Thirring or Bardeen–Petterson effects I mentioned in Chapters 3 and 7 respectively, is not yet clear and more data are required to fully elucidate the observed behaviour. But, there are clues from smaller objects closer to home that may suggest that the precession of jet axes is everything to do with the accretion disc's angular momentum.

Microquasars

The quasars we have been discussing so far are all supermassive black holes that lie at the centres of active galaxies. However, it turns out that there is another class of objects that behave very similarly but are on a much, much smaller scale. These lower mass black holes can be observed rather closer to home, indeed located within our own Milky Way Galaxy, and they are called 'microquasars'. Although the difference in scale size is vast, microquasars in our Galaxy and extragalactic quasars at the centres of other galaxies are both sources of plasma jets with analogous physical properties. Both of these are thought to be powered by the gravitational infall of matter onto a black hole. In the case of a microquasar, the black hole has a mass comparable with that of the Sun. In the case of a powerful extragalactic quasar, the mass of its black hole can be a hundred million times larger than the mass of our Sun. As far as the astrophysicist is concerned, an important advantage of the local examples is that being less massive, they evolve much more rapidly, on timescales of days rather than millions of years in the case of quasars. Nonetheless, as in the case of quasars, the jets which are squirted out from near the centre of all the activity are launched from outside the event horizon, and very likely from the innermost edge of the accretion disc.

Complex mechanisms are at play, and there isn't a simple relationship between the speed at which a jet is launched and the mass of the black hole with which it is associated. In the course of monitoring the jets in the black hole microquasar called Cygnus

X-3 there are occasions when the speeds at which the jet plasma moves away from the black hole are found to vary. This has been measured by time-lapse astronomical measurements in which observations at successive times allow us to determine how fast the jet plasma is hurtling away from the vicinity of the black hole. Such measurements have shown on one occasion the jet speed to be 81% of the speed of light whereas four years later to be 67% of that speed. There is no suggestion that the jet speed is merely reducing with time, since fast and slower jets speeds in this microquasar appear to have been witnessed on a number of occasions since its discovery. Varying jet speeds seem to characterize another well known microquasar in our Galaxy, called SS433, that I shall describe in more detail below. The jet speed in this microquasar seems to change quite a bit as well, indeed it can be anywhere between 20 and 30% of the speed of light over just a few days.

The beauty of symmetry

Figure 23 shows a radio image of SS433, a microquasar in the Galaxy, which is a mere 18,000 light-years distant from us. The striking zigzag/corkscrew pattern is the structure of the plasma jets as they appear to us on the plane of the sky. The individual bolides of plasma that make up the jets are moving at tremendous speeds that vary between 20 and 30% of the speed of light. The directions along which the bolides are moving varies with time in a very persistently periodic way. In fact the axis along which the jets are launched precesses in much the same way as does the paddle of a kayakist, in the frame of reference of the kayak, except on a timescale of six months rather than several seconds. This same behaviour is apparently taking place in at least some quasars (see Figure 21) albeit in that case in such slow motion that we are unable to appropriately time-sample the changes taking place.

The detailed appearance of the zigzag/corkscrew pattern on the sky depends directly on the physical motions of the bolides, as well

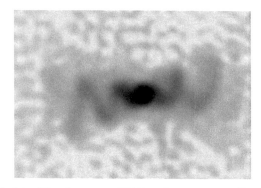

23. The jets of the microquasar SS433 as they appear at radio wavelengths.

as the time when the observation is made. One of the remarkable features of the jets is their symmetry: the physical motions of the components in the eastern jet are equal and opposite to those in the western jet: when one bolide of plasma is at 28% of the speed of light, so too is its counterpart in the oppositely-directed jet; for a different bolide of plasma moving at 22% of the speed of light, so too will its counterpart in the oppositely-directed jet. The fact that one jet appears to have a zigzag structure while the other appears to have a rather different corkscrew pattern is a consequence of the jet plasma always moving at speeds comparable with the speed of light, and well-known relativistic aberrations that occur under such circumstances. The power radiated by this microquasar is rather modest relative to that of an extragalactic quasar but it is still vast in comparison to the power of the Sun which seems somewhat puny, having a total luminosity of only 4×10^{26} watts, a factor of a hundred thousand smaller than that radiated from the microquasar in Figure 23.

Jet launch

The Virgo Cluster is a cluster of well over a thousand galaxies just over fifty million light-years distant from the Milky Way. At its

24. A jet of plasma squirted out at speeds close to that of light, from the supermassive black hole at the heart of the M87 galaxy.

heart is a giant galaxy called M87 (an abbreviation of Messier 87, listed in the catalogue produced by the French astronomer Charles Messier). And, at its heart, is a supermassive black hole whose mass is three billion times that of our Sun. Emanating from this is a strong straight jet, as shown in Figure 24.

This jet is readily visible at optical wavelengths, at radio wavelengths, and at X-ray wavelengths. It is thought that the infalling matter accretes at a rate of two to three Sun's worth of mass per year, onto the very central nucleus where an accretion disc of the sort described in Chapter 6 is thought to be at work. The speed at which this jet propagates away from its launch point, likely at the innermost region of the accretion disc, is very close to the speed of light, and so we refer to it as a relativistic jet. Jet speeds close to the speed of light are revealed by successive monitoring with the VLBA instrument that I introduced in Chapter 7, and the Hubble Space Telescope and Chandra X-ray satellites which are each above Earth's atmosphere and thus attain higher sensitivity than if they were on the ground. At 50 million light-years from Earth an object moving at the speed of light would move across the sky at four milli-arcseconds per year. When we consider that an arcsecond is 1/3600 of a degree, then four-thousandths of this may sound like a tiny angle to measure, but such separations are easily resolvable with an instrument like the VLBA. The VLBA has already imaged the base of this jet to within less than about thirty Schwarzschild radii of its supermassive black hole.

Figure 25 shows an example of the lobes and plumes of radio emitting plasma fed by the relativistic jets from the supermassive black hole in M87.

By way of further illustration that expansive lobes are associated with relativistic jets, Figure 26 shows an example that extends 6 degrees across the sky, and is shown to give a sense of scale with respect to the telescope array used to make the observation. The telescope, used by Ilana Feain and her colleagues, was the Australia Telescope Compact Array.

The mechanisms by which relativistic jets are launched from the vicinity of a black hole remain much closer to conjecture than to acceptance beyond all reasonable doubt. Nonetheless, various independent lines of research by entirely independent teams based in different countries around the world seem to be implying that the preponderance of evidence is that the basic emerging details are correct. Beyond the broad picture, however, the mechanisms and their detailed functioning are conjectural, but being patiently tested amid insufficient photons and selection effects. Proof doesn't belong in science but evidence very much does. We are hindered because even the most advanced imaging techniques deployed today cannot separate and resolve the smallest regions where most of the energy is released, but this is where numerical simulations on powerful computers can transcend the limitations of current technology. Indeed results from simulations of jet launch from accretion discs that fully account for general relativity effects are just being published. These simulations, with known input ingredients and axioms, allow jets and discs to evolve to size-scales where their properties can be confronted against state-of-the-art observations.

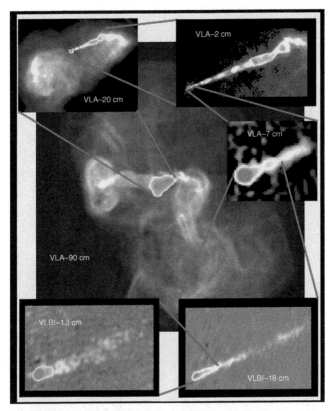

25. The radio-emitting lobes that are fed by the relativistic jet emanating out of the supermassive black hole at the centre of the M87 galaxy.

So what do we now know about the masses of black holes in the Universe? It seems that they fall into two main classes. First, those that have masses similar to those of stars. These stellar mass black holes come in between around three to thirty times the mass of our Sun and come from stars that have burned all their fuel.

26. Composite picture showing an optical image of the moon and the Australia Telescope Compact Array, and a radio image of Centaurus A.

Then there are the supermassive black holes which go all the way up to about ten billion solar masses. As we have discussed, these are found in the centres of galaxies including our own and are responsible for the extraordinary phenomena of active galaxies and quasars.

We have talked about things falling into a black hole, but what happens when a black hole falls into a black hole? This is not an abstract question, since it is known that black hole binaries can exist. In such objects two black holes are in orbit around each other. It is thought that, because of the emission of gravitational radiation, the black holes in a binary will begin to lose energy and spiral into each other. In the final stages of this spiralling, general relativity is pushed to breaking point and the black holes suddenly coalesce into a single black hole with a common event horizon. The energy released in the merger of two supermassive black holes in a binary system is staggering, potentially more than all the light in all the stars in the visible Universe. Most of this energy is dumped into gravitational waves, ripples in the curvature of spacetime, which propagate across the Universe at the speed of light. The hunt is on for evidence of these waves. The idea is that as a gravitational wave passes by a material object, like a long rod, its length will fluctuate up and down as the ripples in spacetime curvature flow through it. If you can measure these tiny length changes, using a technique such as laser interferometry, then you have got a method to detect gravitational waves produced elsewhere in the Universe. Both ground- and space-based gravitational wave detectors, examples of which have been built and more of which are planned, have the potential to pick up signals from black hole mergers. In fact, gravitational waves are so difficult to detect that you need a very strong source to have any chance of such experiments working, and a black hole merger is high on the list of candidates for such strong sources. At the time of writing, gravitational waves have not yet been directly detected, but the experiments are ongoing.

Our best theory of gravity, which comes from Einstein's general theory of relativity, has survived countless tests since its discovery in 1915. It has been shown to give far better agreement to experiment than Newton's theory which it supplanted. However, if general relativity is ever going to be tested up to its limits, you can confidently expect that black holes will prove to be the ultimate testing ground of this cornerstone of modern physics. Where gravity is the most intense in the smallest region of space, so that quantum effects should be important, is exactly where general relativity might break down. However, it might also be that general relativity breaks down on large scales in the Universe. Of course, a hot topic at present is the completeness of general relativity to explain the accelerated expansion of the Universe on the largest scales. Possible deviations away from general relativity are being discussed in connection with accelerated expansion and dark energy. If gravitational waves are detected from the mergers of black holes, or if observations extend our understanding of the fundamental physics which occurs in the vicinity of these fascinating objects, then there's a good chance that we will be able to see how well Einstein's theory holds up or whether it needs to be replaced by something new.

Why do we study black holes?

There are a number of reasons for investigating black holes and one is that they open up the exploration of physics parameter space that is otherwise simply inaccessble to the budgets of even an international consortium. Black hole systems represent the most extreme environments that we can explore, and as such probe the extremes of physics. They bring together both general relativity and quantum physics whose unification has not yet been achieved and remains very much a frontier of physics. A second reason is that trying to understand black hole phenomena arouses fascination in scientists and many thoughtful lay people, and provides a route by which many people can be stimulated by science and motivated to learn about the almighty magnificence of the Universe around us. A third and perhaps surprising reason is Earthly spin-offs. How could black hole research conceivably change our lives? The answer is that it has already done so. As I type these final sentences of this little book into my laptop, it simultaneously backs up my work onto my University server via the 802.11 WiFi protocol. This intricate and clever technology emerged directly out of a search for a particular signature, at radio wavelengths, of exploding black holes led by Ron Ekers to test a model suggested by (now Astronomer Royal, Lord) Martin Rees. Ingenious radio engineers in Australia, led by John O'Sullivan, in the course of devising interference suppression algorithms for the tricky business of detecting subtle signals from distant space realized that these could be applied to transform communication here on Earth. Black holes therefore have the power to rewrite physics, reinvigorate our imagination and even revolutionize our technology. There are many spin-offs from black holes—way beyond their event horizons.

Further reading

M. Begelman and M. Rees, *Gravity's Fatal Attraction*, 2nd ed. (Cambridge University Press, 2010).

J. Binney, *Astrophysics: A Very Short Introduction* (Oxford University Press, 2015).

J. B. Hartle, *Gravity* (Addison Wesley, 2003).

A. King, *Stars: A Very Short Introduction* (Oxford University Press, 2012).

A. Liddle, *An Introduction to Modern Cosmology*, 3rd edn. (Wiley-Blackwell, 2015).

F. Melia, *The Galactic Supermassive Black Hole* (Princeton University Press, 2007).

D. Raine and E. Thomas, *Black Holes: An Introduction*, 2nd edn. (Imperial College Press, 2010).

C. Scarf, *Gravity's Engines: How Bubble-Blowing Black Holes Rule Galaxies, Stars, and Life in the Cosmos* (Scientific American/ Farrar, Straus and Giroux; Reprint edition, 2013).

R. Stannard, *Relativity: A Very Short Introduction* (Oxford University Press, 2008).

A. Steane, *The Wonderful World of Relativity: A Precise Guide for the General Reader* (Oxford University Press, 2011).

K. Thorne, *Black Holes and Time Warps* (W. W. Norton, 1994).

Index